Handbook of Offshore Helicopter Transport Safety

Woodhead Publishing in Mechanical Engineering

Handbook of Offshore Helicopter Transport Safety

Essentials of Underwater Egress and Survival

Edited by

MICHAEL J. TABER

ELSEVIER

Amsterdam • Boston • Cambridge • Heidelberg
London • New York • Oxford • Paris • San Diego
San Francisco • Singapore • Sydney • Tokyo
Woodhead Publishing is an imprint of Elsevier

WP WOODHEAD PUBLISHING

Woodhead Publishing is an imprint of Elsevier
80 High Street, Sawston, Cambridge, CB22 3HJ, UK
225 Wyman Street, Waltham, MA 02451, USA
Langford Lane, Kidlington, OX5 1GB, UK

ISBN: 978-1-78242-187-0 (print)
ISBN: 978-1-78242-188-7 (online)

British Library Cataloguing-in-Publication Data
A catalogue record for this book is available from the British Library

Library of Congress Control Number: 2015944128

For information on all Woodhead Publishing publications visit our website at http://store.elsevier.com/

CONTENTS

ABOUT THE EDITOR

Michael J. Taber, MSc, PhD

Dr Michael J. Taber is the Senior Research Scientist at Falck Safety Services Canada. He specializes in human performance in harsh environments. After his time as an aviation technician and safety diver in the Royal Canadian Air Force, Dr Taber worked as the Senior Offshore/Military HUET instructor for an internationally recognized training provider. Michael has completed an undergraduate degree in psychology, a Master of Science (kinesiology), an interdisciplinary PhD (psychology/human factors/industrial engineering), and two postdoctoral fellowships. He has lectured at two universities in three departments and currently holds Adjunct Professor status at Brock University (Kinesiology Department) and Memorial University (School of Human Kinetics and Recreation). Dr Taber has published over 75 scientific papers, reports, and chapters exploring emergency response, user experience, HUET-related performance, and factors affecting survival.

ABOUT THE AUTHORS

Martin Barwood, Fellowship of the Higher Education Academy **(FHEA), BSc, PhD**
Martin Barwood is a Research Fellow at Northumbria University in the Department of Sport, Exercise and Rehabilitation. He received his interdisciplinary PhD in Environmental Physiology and Psychology from the University of Portsmouth in 2006. Since graduating, Dr Barwood has accumulated nearly 10 years of research experience investigating the acute, short- and long-term physiological and psychological responses to cold-water immersion. He has also contributed to knowledge on adaptation to cold water. He is an active member of the open-water safety community and has conjointly published more than 100 academic papers, reports, and conference abstracts.

Sue Coleshaw, PhD, BSc, Chartered Ergonomist & Human Factors Specialist **(C.ErgHF)**, Fellow of the Institute of Ergonomics and Human Factors **(FIEHF)**
Dr Sue Coleshaw is an Independent Research Consultant based near Aberdeen, UK. She has a BSc from the University of Leeds and a PhD from the University of London in applied thermal physiology. She is a Fellow of the Institute of Ergonomics and Human Factors and is a Chartered Ergonomist and Human Factors Specialist. Providing advice to legislators, equipment manufacturers, and the offshore industry, her specialist interests include helicopter safety, human performance, offshore emergency response, life-saving, and personal protective equipment. Sue Coleshaw has published numerous academic papers, industry reports, and conference contributions. She is actively involved in the work of a number of international standards committees and is the author of a technical standard for helicopter emergency breathing systems (EBS).

Sean Fitzpatrick
Sean Fitzpatrick is currently the Operations Manager for Falck Safety Services Canada's Newfoundland training center. He is an experienced aircraft ditching and sea survival instructor and instructor trainer for civilian aviators, offshore oil and gas workers, and military aircrew (both helicopters and fixed-wing aircraft). Training includes the use of emergency breathing systems, awareness of the effects of sudden cold-water immersion, and

mission-specific training development and delivery. Sean also provides international training including HUET Diver and Instructor Development courses, assisting training centers with attaining approvals, and providing customized on-site courseware and program delivery. To date, he has worked with civilian and military groups in Canada, Australia, France, Scotland, Russia, USA, Norway, Egypt, South Korea, and India.

Sean spent 8 years as a member of the Canadian Forces. He was employed at various bases in both Canada and Europe and served as a United Nations Peacekeeper in Croatia in 1993. He worked as a professional scuba diver and dive instructor. Over a 10-year span, his diving duties took him across Canada and throughout the Caribbean both on and under water. He remains an avid diver.

John Leach, PhD

Dr John Leach is a research psychologist specializing in the psychology of human survival and author of the book *Survival Psychology* (Macmillan, 1994). He was lecturer in cognitive psychology at the University of Lancaster (England) for 16 years where he delivered courses on psychology in extreme environments and cognitive evolution. He is also trained as a Royal Air Force Combat Survival (SERE) officer and qualified as an instructor in sea, desert, jungle, arctic, and polar survival and in conduct after capture. While working as a SERE psychologist, he developed a cognitive theory of human behavior in survival situations (dysexecutive survivor syndrome) that won the 2013 William E. Collins Award from the Aerospace Human Factors Association. He has completed various military and civilian HUET courses. John Leach is an affiliated member of the Center for the Study of Human Cognition at the University of Oslo, Norway.

Sarita J. Robinson, MSc, PhD

Sarita J. Robinson is an Associate Professor with over 15 years of research experience investigating how people respond in life-threatening situations, using a cognitive neuropsychology and psychobiology approach. After completing her Master's degree in Cognitive Sciences at the University of Manchester, Sarita went on to complete her PhD in cognitive and neuroimmune changes in human survival. Since the completion of her PhD in 2004, Sarita has been based at the University of Central Lancashire, United Kingdom (UK), and has continued her research in the survival field working with both UK-based and international clients.

Dana H. Sweeney, MSc

Dana H. Sweeney is a Research Scientist and offshore survival/HUET/TEMPSC/FRC instructor at Falck Safety Services Canada (FSSC). He holds a Bachelor's degree in Kinesiology and a Master's degree in Environmental Physiology. Dana was a coxswain with the Canadian Coast Guard Inshore Rescue Boat program, which offered experience with life-saving equipment, rough-weather situations, public relations work, and boating safety advocacy through search and rescue operations and training. Dana's Masters research was related to hypothermia, motion sickness, and survival at sea and was completed under a Defence Research and Development Canada Research Fellowship in 2003.

Since 2003, Dana has been involved with the manufacturing of HUET simulators; research, evaluation and approval of immersion suits; helicopter passenger transportation suits; personal flotation devices (PFDs); lifejackets; life rafts; and protective clothing; using humans, thermal manikins, and flotation manikins. Dana has worked for clients such as Transport Canada, Department of National Defence, National Research Council of Canada, Canadian Association of Petroleum Producers, and numerous equipment manufacturers. He has also been involved in test method development for survival equipment standards with Canadian General Standards Board and Underwriters Laboratories. Dana now contributes his applied and academic expertise to the research questions with Falck Safety Services Canada.

Michael J. Tipton, MSc, PhD

Michael J. Tipton is a Professor of Human and Applied Physiology at the Extreme Environments Laboratory, Department of Sport and Exercise Science, University of Portsmouth. He has spent over 30 years researching and advising the military, industry, and elite sports people in the areas of thermoregulation, environmental and occupational physiology, and survival in the sea. He has published over 500 scientific papers, reports, chapters, and books in these areas. Professor Tipton sits on the Royal National Lifeboat Institution's Medical and Survival Committee and is a Trustee of Surf Life Saving Great Britain (GB), he was made an Honorary Life Member of the International Association for Safety and Survival Training in recognition of his work in sea survival.

PREFACE

Traveling to and from an offshore installation or vessel by helicopter is part of the everyday routine for thousands of individuals around the world. For the most part, the trip is uneventful and often becomes a mundane task that is just part of working offshore. On rare occasions, this trip could turn into anything but normal or routine. It is during these rare emergency events that we need to ensure individuals have the necessary skills to complete a complex set of tasks in what is often an extremely short period of time. The preparatory training to egress from a ditched, possibly capsized, flooded, and/or sinking helicopter is only one part (albeit a critical component) of the overall survival situation.

This book explores the myriad of factors that need to be considered when preparing for a helicopter ditching or water impact. From a historical perspective, the development and implementation of a helicopter underwater egress training (HUET) program is outlined to identify specific aspects that influence skill acquisition and retention of the procedural skills. The design and use of HUET simulators are identified as key components to ensuring that individuals have an opportunity to practice egress skills in a safe, realistic environment. However, the ability to make decisions while in a stressful situation is also considered a critical component of the survival of personnel in this often lethal set of circumstances.

This book represents a dedicated commitment to ensuring the safety of offshore personnel who operate over some of the most hostile places on the planet. World-leading authorities in human physiology associated with immersion in cold water and habituation techniques, survival psychology associated with decision-making/executive function, performance while under stress, development and use of survival equipment (emergency breathing systems and immersion suits), and HUET program factors have combined their knowledge to provide the reader with a comprehensive reference that can be used to make informed decisions about how to best prepare the global offshore workforce to survive a helicopter ditching/water impact. It is our hope that the information provided in this book will aid in the preparation of these individuals and ultimately increase the overall chances of survival.

CHAPTER 1

Introduction to Factors Influencing Offshore Helicopter Transport Safety

Michael J. Taber
Falck Safety Services Canada Inc., Dartmouth, NS, Canada

1.1 WHY A BOOK ON OFFSHORE HELICOPTER SAFETY?

First and foremost, this book was written with the primary goal of informing and preparing the millions of individuals who travel over water in a helicopter. Whether you hold the flight controls, maintain the many operational systems, or are a passenger during these tens of millions of flights, the information contained in this book is designed to offer a detailed collection of information that could better prepare personnel to respond if something goes wrong and the helicopter crew needs to make a decision to land on the surface of the water. It is my hope that, armed with this information, you will consider the variety of ways in which you can actively be engaged in your own safety during these flights.

To provide some context to the number of overwater flights that occur in the oil and gas industry, the Helicopter Safety Advisory Conference (2014) reported that an average of 2,307,193 passengers were flown to installations in the Gulf of Mexico between 2009 and 2013. Oil & Gas UK (2011) reported that between 1981 and 2010 helicopters on the UK Continental Shelf transported roughly 54,000,000 offshore personnel. Not even during military transport in times of war do we witness this magnitude of moving individuals by helicopter to their place of work. Offshore helicopter operations are one of only a few organized/regulated occupationally required forms of travel for millions of people every year. With so many individuals using this mode of transportation, the numbers of regulators, the different types of helicopters, and the diverse environments in which operations are carried out, it is important to consider the factors that influence our collective understanding of this unique aspect of offshore life.

This book addresses offshore helicopter operations from a safety perspective with a primary focus on helicopter underwater egress training

Handbook of Offshore Helicopter Transport Safety
ISBN 978-1-78242-187-0
http://dx.doi.org/10.1016/B978-1-78242-187-0.00001-8

(HUET). By encompassing issues directly related to offshore helicopter transport safety such as safety management systems (SMS), current airframe usage, preflight preparation, safety briefings, survival equipment, underwater egress training, water impact/ditching statistics, and search and rescue, it is my aim to better prepare individuals for an event that, since 2000, has on average claimed the lives of approximately 30% of those involved. Each area of interest will detail pertinent information spanning approximately 30 years of offshore operations. The goal of the information presented here is to provide a comprehensive reference guide for offshore operators, oil and gas operators, safety and survival instructors, and the global offshore workforce, as well as anyone with an interest in overwater helicopter operations.

From this perspective it is possible to explore the early years of HUET as part of a recognized necessity during the latter part of World War I (WWI) through to the present day training requirements. From the very first reported ditching event in 1944 (Brooks, 2007) to the more recent event off the Shetland Islands in 2014 (AAIB, 2014), there are commonalities in the factors affecting survival that have yet to be fully addressed by regulators, training organizations, and operators. In most cases, the factors affecting survival are fairly straightforward; in others there is a complex combination of issues that, as an industry, we may never be able to fully mitigate. However, by including contributions from some of the world's leading authorities on survival, it is my goal to shed light on the problems that need to be overcome, as well as to offer some possible solutions.

Early ditching experiences outlined in accident investigation reports appear to have focused primarily on human factors related to performance in extreme situations. This focus was firmly grounded in a systems approach that considered only the interactions between the pilot and the surrounding environment. These initial investigations suggested that engineering the problems out of the system was believed to be the best possible solution, as humans were thought to be fallible and would inevitably make mistakes (Dekker, 2014). By creating systems that were resistant to the influences of human errors, it was possible to limit the number of failure events as well as the associated consequences. Unfortunately, a systems approach cannot address issues related to how processes can be worked around if someone is motivated enough or uses the equipment in a manner that it was never intended for (i.e., more error resistant than error proof).

During WWII, human factors in engineering and personnel selection had been refined to the point that aviation errors associated with the pilots and crew were greatly diminished (Dekker, 2005; Strauch, 2007). However, occasional errors still occurred and catastrophic events remained (and still remain) part of the culture for those who make their living behind the controls of an aircraft. Investigations of these events typically focus on the problems from a mechanistic point of view. Not surprisingly, the reports often contain exhaustive detail outlining the intricate working of a mechanical component that was suspected to have caused the problem. If no mechanical failure could be found, there was an assumption that the event had to be caused by human error (Dekker, 2014; Reason, 1990).

Of particular interest to this book, human survival aspects related to a ditching or water impact tend to occupy a relatively small amount of the lengthy investigation reports (Brooks, MacDonald, Donati, & Taber, 2008; Taber & McCabe, 2006, 2009). Even in cases in which the primary or root cause of the event is identified as "human error," reports often leave out critical survival aspects such as whether the individuals completed previous egress training, what type of training was completed, the type of simulator used during training, water temperature, wave height, whether emergency breathing systems were available, or if they were used (Brooks et al., 2008). Additionally, there is not typically an overview of factors related to caffeine levels, amount of sleep in the 12 h prior to the event, hydration, caloric intake, or ability to cope with a stressful situation of the pilot or crew. Without this critical information, it is unlikely that training programs designed to prepare individuals for the difficulties associated with a helicopter ditching/water impact will change to be more representative of the actual situation. By providing the specific information related to egress or postegress survival, rather than indicating only that the individual drowned or was unable to escape, it will be possible to provide guidance to HUET program developers and instructors.

To address the factors associated with development of a better HUET program and ultimately to increase survivability, this book begins by exploring the pioneers who believed that training individuals to egress from a capsized and flooded helicopter was a fundamental responsibility for those who asked them to fly. By exploring the early beginnings, it will be identified that the focus of egress training initially followed a systems approach (Reason, 2000) that examined the aircraft design, exit placement and size requirements, and type/position of survival equipment. Later this focus began to address the issues from a person approach (Reason, 2000). However, this approach suggests that errors are the result of shortcomings in attention, focus, and motivation that lead to forgetfulness, carelessness, neglect, and recklessness (Burnham, 2009; Reason, 2000).

From a person approach, it is difficult to consider the problems or errors related to physiological and psychological aspects of those who had to egress from a capsized and flooded cabin in such a way that appropriate mitigation strategies can be implemented. The difficulty with developing this approach lies in the fact that the event occurs rapidly (Brooks et al., 2008), is limited to first-hand accounts of those who survive, and occurs in a water environment. Limiting factors to this approach stem from the fact that it is difficult to piece together the reasons why some individuals survive a ditching (despite having never completed a HUET program) and why some do not (despite having been trained and uninjured during the event). However, if the focus was to include both a systems and a person approach during investigations, HUET programs could begin to address the multitude of factors affecting optimal performance from a holistic perspective known as human systems integration (HSI).

1.1.1 Human Systems Integration

While reading this book, it is important to consider the influencing factors associated with offshore helicopter transport from the perspective of how to best prepare individuals for an emergency event. Although the day-to-day operations are an important component of the overall process, this book explores the process of training in underwater egress and survival. This examination is carried out from a human systems interaction perspective to ensure that all factors (from the moment that a training need is required to the moment that someone has survived and been rescued after an event) are considered in such a manner that survivability is improved.

First developed by US Army researchers (Booher, 2003), an HSI (originally known as MANPRINT) framework was designed to systematically investigate human factor issues such as personnel selection and equipment design criteria in an attempt to decrease errors and increase mission success. HSI now includes 13 different components that range from human factors engineering to assessments of survivability. By combining a systematic review of the environmental conditions in which a task will be performed, the available technologies that could be used in that environment, and the human capabilities/limitations that influence performance of the task, it is possible to identify an optimal training and evaluation process to increase retention of information, usability, habitability, and survivability. Without including the three primary components (environment, technology, and humans) in a performance/outcome review process, emphasis can be (and often is) placed in the wrong area.

For example, a typical examination of helicopter ditching/water impact data focuses heavily on the cause of the event with minimal information that could be used to better prepare the majority of individuals who are not flying the aircraft.

1.1.2 Helicopter Accident Statistics

To contextualize the HSI approach, several chapters in this book discuss helicopter ditchings or water impacts. The distinction between a ditching and a water impact is important only insofar as the forces that are involved during the initial contact with the water. Obviously, if the forces are so great that they exceed human tolerance level, the likelihood of survival is extremely limited. Typically, these types of events are identified as water impacts and occur less frequently than a semicontrolled landing. Conversely, an event that includes limited impact forces while the helicopter is still under the control of the aircrew is most often termed a ditching. Ditchings, particularly events in which the helicopter remains upright on the surface, represent a greater chance of survival than a water impact. Research has also shown that survival rates are higher if the helicopter remains at the surface, regardless of whether it is upright or inverted (Taber & McCabe, 2007).

Interestingly, the Civil Aviation Authority (CAA) (2013) provides a definition of a nonsurvivable event as a situation in which there are "no survivors or only a very small number of the occupants survived" (p. 30). Based on the distinction between a ditching and a water impact, this definition of a nonsurvivable event is difficult to reconcile for events in which some of the individuals survive. Therefore, for the purpose of this book, no distinction between ditching and water impact is made. A survivable event is defined as one in which at least one individual egressed and was rescued. A nonsurvivable event is defined as one in which the impact forces were sufficient enough to render all individuals unconscious (thus unable to hold their breath) and/or all individuals received lethal injuries during the impact.

To explore the prevalence of helicopter accidents, this book explores a number of events in an effort to identify areas of improvement regarding overall survival. For example, in the CAA (2013) report it was noted in the review of past occurrences that seven of the 28 events (25%) that occurred between 1967 and 2012 were nonsurvivable water impacts. The seven events represented ~44% of the 16 water impacts; however, it was not identified if anyone survived the nonsurvivable events (CAA, 2014). Intuitively, it could be assumed that no one survived; however, given the definition provided by the CAA, it is not entirely clear. For example, if the CAA criteria of nonsurvivable events were applied to events such as those that occurred on March 12, 2009 (Cougar 491), it is possible that the ditching would not have been included in a similar review (only one survivor). However, the accident investigation report clearly identified that all passengers and crew survived the initial impact forces (TSBC, 2010). Without including all ditching events in a review, it is difficult to have a true understanding of what areas (if any) could be improved for future operations.

1.1.3 Relevant Safety Legislation and Current Standards

Relevant legislation and safety standards are considered from an HSI framework to ensure that aspects of personnel selection, safety training, and hazard identification are included in the analysis of how to best prepare the offshore workforce for emergency situations such as a ditching/water impact. By exploring the relevant safety legislation from the HSI framework it is possible to identify gaps in the specific jurisdictional requirements. Through the identification of these gaps it might be possible to develop a universal standard or at least improve the reciprocity of training recognition on a global scale.

1.1.4 Safety Culture versus Safety Climate

Transport Canada (2008) defines safety culture as "the product of individual and group values, attitudes, perceptions, competencies and patterns of behaviour, that determine the commitment to, and the style and proficiency of, an organization's health and safety management" (Section 3.5.2). Taber (2010) writes, "Taken together, SMS, climate, and culture represent the collective aspects of the offshore workforce, the pilots that transport them to their destinations, the organizational structure in which they work, the environmental factors of their geographical location, and the technology of the equipment used to keep the system operating in an effective, efficient, and safe manner" (p. 47). James Reason (1997) suggests that, "an ideal safety culture is the engine that continues to propel the system towards the goal of maximum safety health, regardless of the leadership's personality or current commercial concerns" (p. 195).

Safety culture in the oil and gas industry has a long and well-documented history. Safety climate on the other hand is rarely discussed in formal written information. It has been purported that safety climate (as the name suggests) is sensitive to the changes that occur within an organization and reflects the current mood of the workforce as it pertains to recent safety-related events

(Cox & Flin, 1998). For example, Mearns, Whitaker, and Flin (2003) report that safety climate is "a manifestation of safety culture in the behaviour and expressed attitude of employees" (p. 642). This effect could be seen immediately following the Canadian Helicopters Shetland Island and Cougar 491 ditchings. Results from both the North Sea and the Canadian offshore worker survey questions exploring helicopter safety suggest that 44% and 28% (respectively) of those surveyed felt unsafe when traveling in a helicopter to and from the installation (Wells, 2010). The formal identification of safety concerns does not necessarily indicate that the safety culture or SMS of the organization is deficient; it does, however, identify that a change in observable workforce behavior is reactive to recent events and, as Reason (1997) points out, is often "short-lived" (p. 192).

1.1.5 Overview of Focused Chapters

This final section provides an overview of the focused chapters and highlights the diverse nature of the factors that influence survival after a helicopter ditching/water impact event. The chapters cover a wide range of topics from the development of HUET programs to the physiological and psychological preparation/responses associated with performing egress-related survival tasks. Each chapter is designed to increase or consolidate our collective understanding of the factors associated with the essentials of underwater egress and survival.

Chapter 2 (Mr Sean Fitzpatrick) explores the development and implementation of HUET programs. A short vignette describing the experience of someone who is involved in a ditching is used to contextualize the presented information. This information forms a reference point for the differences in training methodologies as well as provides an overview of how the programs have progressed over the past 30+ years. From this starting point it will be possible to identify where improvements can be made with respect to existing technologies and where advances in applied training technologies may still be needed.

Chapter 3 (Dr John Leach) provides an examination of the psychological factors associated with survival in general terms. The chapter identifies critical aspects of decision-making in an effort to aid in the formulation of a survival plan. Particular attention is paid to the cognitive malfunctions in working memory, executive attention, and the supervisory system. This information is critical when considering what information is most important during the initial skill-acquisition process to ensure a greater level of retention.

Chapter 4 (Dr Michael J. Tipton) considers survival from a more physiological standpoint while remaining general in terms of overall survival. Specifically, the chapter explores human responses to sudden cold-water immersion. The four phases of cold-water immersion are used to frame a consideration for the implementation of supplemental emergency breathing systems that could aid individuals in a ditching situation.

Chapter 5 (Dr Martin Barwood) begins the process of integrating several aspects of psychology and physiology from a more HUET-focused perspective. Psychological preparation for an emergency event is considered in regard to typical physiological responses and offers insight into possible mitigation strategies that could be employed in both a short- and a long-term approach. The chapter consolidates several cold-water breath-hold research projects to identify how individuals could better prepare for the physiological effects associated with sudden immersion.

Chapter 6 (Dr Sarita J. Robinson) furthers the discussion of psychological effects related specifically to HUET program performance. The information considers how changes in neurochemical balances in the brain could lead to cognitive failures while under stress. Based on previous research, recommendations are suggested for ways to better prepare for HUET performance requirements and ultimately for survival in an actual event.

Chapter 7 (Dr Michael J. Taber) explores the implications of using various levels of fidelity in skill acquisition and retention. A distinction is made between physical, functional, and cognitive/psychological fidelity. By exploring issues such as cabin configuration, problems associated with crash-attenuation seats, and proximity to an emergency exit, it is posited that an integration of contextual interference in training may improve retention of HUET skills. An argument is also presented for why there is a need to ensure a high level of physical fidelity during training and why the HUET program is an ideal setting to develop acute stress management techniques that can be implemented in the event of an emergency.

Building on the inclusion of HUET program training equipment and course material, Chapter 8 (Dr Sue Coleshaw) discusses the use of emergency breathing systems (EBSs). The chapter provides a detailed overview of the two specific types of EBS currently used in HUET programs. A discussion of various design criteria identifies why there has been a shift from the rebreather-style EBS to compressed-air EBS in the UK oil and gas industry. The chapter also presents information regarding training techniques that need to be considered to address acute stress within the HUET program.

Chapter 9 (Mr Dana H. Sweeney) continues the examination of HUET program training and operational equipment. Specifically, this chapter provides detailed information about helicopter transportation suits (HTSs) and associated life jackets. Given the large number of individuals involved in offshore operations in areas such as Canada, Norway, England, Scotland, and Holland, the perspective of the HTS is taken from a cold-water environment. Particular attention is paid to issues such as fitting, water ingress, and supplementary survival equipment to ensure the highest possible level of survival in the event of a ditching/water impact.

Chapter 10 (Dr Michael J. Taber) presents a consolidated summary of the information contained in each chapter. The consolidation of the information presented in this book represents the collective knowledge of some of the world's leading authorities on HUET-related training. To capitalize on this knowledge and provide some guidance for future improvements, this summary is used to present an optimal HUET program curriculum that could be used internationally.

REFERENCES

Air Accident Investigation Branch. (2014). Special bulletin S1/2014. AS332 L2 Super Puma, G-WNSB.

Booher, H. R. (Ed.). (2003). *Handbook of human systems integration*. New Jersey: Wiley.

Brooks, C. J. (2007). The human factors of surviving a helicopter ditching. In *Research and Technology Organization, North Atlantic Treaty Organization, Survival at sea for mariners, aviators, and search and rescue personnel*. RTO AGARDograph, AG-HFM-152.

Brooks, C. J., MacDonald, C. V., Donati, L., & Taber, M. J. (2008). Civilian helicopter accidents into water: analysis of 46 cases, 1979–2006. *Aviation, Space, and Environmental Medicine, 79*(10), 935–940.

Burnham, J. C. (2009). *A history of technology, psychology, and misfits of the machine age: Accident prone*. Chicago: The University of Chicago Press.

Civil Aviation Authority. (2013). *CAP 1034 – Development of a technical standard for emergency breathing systems*. Norwich, UK: TSO (The Stationary Office) on behalf of the UK Civil Aviation Authority.

Civil Aviation Authority. (2014). *CAP 1145 – Safety review of offshore public transport helicopter operations in support of the exploitation of oil and gas*. West Sussex, UK: Safety and Airspace Regulation Group.

Cox, S., & Flin, R. (1998). Safety culture: philosopher's stone or man of straw. *Work and Stress, 12*(3), 189–201.

Dekker, S. (2014). *A field guide to understanding human error* (3rd ed.). Burlington, VT: Ashgate Publishing Company.

Dekker, S. W. A. (2005). *Ten questions about human error: A new view of human factors and system safety*. Mahwah, New Jersey: Lawrence Erlbaum Associates, Publishers.

Helicopter Safety Advisory Conference (HSAC). (2014). *Helicopter safety advisory conference (HSAC): 2013 Gulf of Mexico offshore helicopter operations*. Available from http://www.hsac.org/portals/45/HSAC%202013%20Report.pdf.

Mearns, K., Whitaker, S. M., & Flin, R. (2003). Safety climate, safety management practices and safety performance in offshore environments. *Safety Science, 41*, 641–680.
Oil & Gas UK. (2011). *UK offshore commercial air transport helicopter safety record (1981–2010)*. The United Kingdom Offshore Oil and Gas Industry Association trading as Oil & Gas UK. Document available at http://www.oilandgasuk.co.uk/cmsfiles/modules/publications/pdfs/HS027.pdf.
Reason, J. (1990). *Human error*. New York: Cambridge University Press.
Reason, J. (1997). *Managing the risks of organizational accidents*. Burlington, VT: Ashgate Publishing Limited.
Reason, J. (2000). Human error: models and management. *British Medical Journal, 320*(7237), 768–770.
Strauch, B. (2007). *Investigating human error: Incidents, accidents, and complex systems*. Burlington, VT: Ashgate Publishing Company.
Taber, M. J. (2010). Offshore helicopter safety inquiry report. In Wells, R. W. (Commissioner) (Ed.), *Canada-Newfoundland and Labrador offshore helicopter safety inquiry. Vol. II*, pp. 211–290. Document available at http://oshsi.nl.ca/?Content=Reports.
Taber, M. J., & McCabe, J. (2006). Helicopter ditching: time of crash and survivability. *SAFE Journal, 34*(1), 5–10.
Taber, M. J., & McCabe, J. (2007). An examination of survival rates based on external flotation devices: a helicopter ditching review from 1971 to 2005. *SAFE Journal, 35*(1), 1–6.
Taber, M. J., & McCabe, J. (2009). The effect of emergency breathing systems during helicopter underwater escape training for land force troops. *Safety Science, 47*(8), 1129–1138.
Transport Canada. (2008). *Guidance on safety management systems development*. Advisory Circular 107-001 Issue 01.
Transportation Safety Board of Canada. (2010). *Main gearbox malfunction/collision with water: Cougar Helicopters Inc., Sikorsky S-92A, C-GZCH, St. John's Newfoundland and Labrador, 35NM E, 12 March 2009*. Aviation Investigation Report A09A0016. Minister of Public Works and Government Services Canada.
Wells, R. (2010). *Canada-Newfoundland and Labrador offshore helicopter safety inquiry* (Vol. I). St. John's, NL: Canada-Newfoundland and Labrador Offshore Petroleum Board.

CHAPTER 2

Development and Implementation of Helicopter Underwater Egress Training Programs

Sean Fitzpatrick
Falck Safety Services Canada Inc., Dartmouth, NS, Canada

2.1 INTRODUCTION

What makes one helicopter underwater egress training (HUET) program better than another? What is going through the mind of a helicopter passenger about to ditch into icy cold water? What information and skill sets do trainees need to survive a real-world ditching event? How should HUET programs be organized and delivered? Does the trainees' belief that HUET does or does not matter contribute to their survivability, and are training organizations evaluating their performance appropriately? What characteristics are essential for creating effective HUET programs?

The information contained in this chapter is intended for training providers, standards organizations, and HUET instructional teams as well as those who travel by helicopter over water to and from offshore installations. It discusses the history of ditching training from pre-World War I to the present day. The information is contextualized by providing a hypothetical first-hand account of what a ditching survivor might have to face. From this perspective, the information provides an overview of the various HUET programs currently being delivered around the world. Four phases of a ditching are also described in an effort to outline vital information for each phase. The chapter discusses safety protocols and provides insight into how training programs should be organized and delivered. Finally, a framework outlines the necessary training outcomes, performance evaluation, and characteristics of an effective HUET program.

2.1.1 History of Ditching Training

HUET programs and simulators have been developing and evolving since the first instances of aircraft ditchings during takeoff or landing from aircraft carriers in the early 1900s. One of the earliest known ditchings took place

Handbook of Offshore Helicopter Transport Safety
ISBN 978-1-78242-187-0
http://dx.doi.org/10.1016/B978-1-78242-187-0.00002-X

on July 31, 1912. Lieutenant Theodore G. Ellyson (the US Navy's first aviator) ditched during experiments with an early catapult system at Annapolis, Maryland (http://www.nationalaviation.org/ellyson-theodore/). The A-1 Triad that Ellyson was flying was caught in a crosswind, which sent him and the airplane plunging into the Severn River. At the time, the US Navy focused on equipment and procedures designed to facilitate the recovery of the aircraft if it ditched into the water. There was little emphasis placed on techniques needed to assist aircrew or passengers in escaping from a flooded and, potentially, sinking aircraft.

The US Naval Aviation Museum (n.d.) reports that during World War II, several F4F Wildcat fighters flying from the carrier *Hornet* (CV 8) were forced to ditch as a result of fuel depletion following a mission (http://www.navalaviationmuseum.org/history-up-close/objects-of-history/birth-dilbert-dunker/). Lieutenant John Magda wrote the following in an after-action report: "There should be a landing in water 'check-off list' in every plane, because at a time like that there are a few things you may forget that may prove to be a very dear mistake. There is very little time to do anything after the plane hits the water—thirty seconds at the most" (US Naval Aviation Museum, n.d.). Another VF-8 pilot, Lieutenant H.L. Tallman, wrote, "Shock of landing is not bad, but the water that gushes into the cockpit and the splash caused by the impact leads one to believe at the moment that the plane is going right on down. Actually, by the time you've recovered your senses (1–2 seconds) the water is up to your neck" (US Naval Aviation Museum, n.d.).

Based on these early experiences, HUET programs, as we know them today, originated using the Dilbert Dunker. The Dilbert Dunker (Figure 2.1), which was engineered and built by Ensign Wilfred Kaneb, provided seating for a single pilot. When the simulated cockpit was released, it traveled down a pair of steel rails as fast as gravity would take it and then rolled nose-over as it impacted the water. The pilot was then required to overcome in-rushing water and disorientation and make his/her escape.

Kaneb recalled that one senior officer's recommendation during the Dilbert Dunker's development was to design a machine that would "teach them what it is like to be drowning" (US Naval Aviation Museum, n.d.). In an overview of the Dilbert Dunker history, the US Naval Aviation Museum (n.d.) posits that Kaneb's objective was not to teach trainees what it was like to be drowning, rather he wanted to teach them how to orient themselves under water and make a successful escape, or egress, to the surface.

Figure 2.1 Dilbert Dunker at 12 Wing Shearwater c. 1980. *(From Taber and Bohemier (2014) with permission.)*

As the Dilbert Dunker was the only underwater egress simulator (UES) in existence at the time, both helicopter and fixed-wing aircrew personnel used it for training. One of the primary training issues, however, was the fact that the Dilbert Dunker did not roll onto its side as a helicopter does when it ditches into the water. To address this issue, as well as to accommodate multiple trainees at once, the Burtech 9D5 helicopter UES was introduced in 1974 (Figure 2.2). The 9D5 was created to be a simulator that would familiarize aircrew and passengers with escape techniques from a ditched helicopter. The 9D5 used a rudimentary cockpit and generic back end, which could accommodate up to two trainees in the front and four in

Figure 2.2 US Navy Burtech 9D5, NAS, Pensacola, Florida. *(Image courtesy of Smith, Ray E. Naval Aviation Survival Training Program, U.S. Navy.)*

the back (see Cunningham, 1978 for a future overview of the 9D5). It was not intended to simulate any particular helicopter; instead it was considered to be a composite of all US Navy and Marine Corps helicopters being operated at the time (see Chapter 7 for a discussion on generic UESs). This was partially accomplished by the use of a universal-type exit release mechanism (i.e., pull up, pull down, push forward, pull back, etc.).

Following the 9D5, the McLean and Gibson UES was introduced in 1985. Like the 9D5 the McLean and Gibson was designed for use by helicopter aircrews and passengers, rolled onto its side, and included a replicated cockpit on both the front and the back of the simulator, which allowed up to four pilots to train at a time in addition to passengers or flight crews in the cabin area (Figure 2.3). Although an improvement on the 9D5, the McLean and Gibson initially lacked exits (they were added later); also the round shape of the simulator versus the flat sides of a helicopter reduced fidelity (Summers, 1996; see also Chapter 7), and the seats were not configurable.

By the mid-1980s, HUET was becoming common for the oil and gas industry, the military, and civilian aviators around the world and resulted in the demand for a more advanced UES. In 1986, Survival Systems Ltd began construction on a completely new series of Modular Egress Training Simulators (METS™). The first METS™ Model 30 went into service in Dartmouth, Nova Scotia, in 1987 and was used to train both Canadian military aircrews and offshore oil and gas workers (Figure 2.4). The METS™ has continued to evolve and is now capable of replicating dozens of helicopter types, amphibious vehicles, fighter jets, fixed-wing aircraft, and

Figure 2.3 Maclean and Gibson simulator in use to train Royal Canadian Air Force personnel in 1984. *(From Taber and Bohemier (2014) with permission.)*

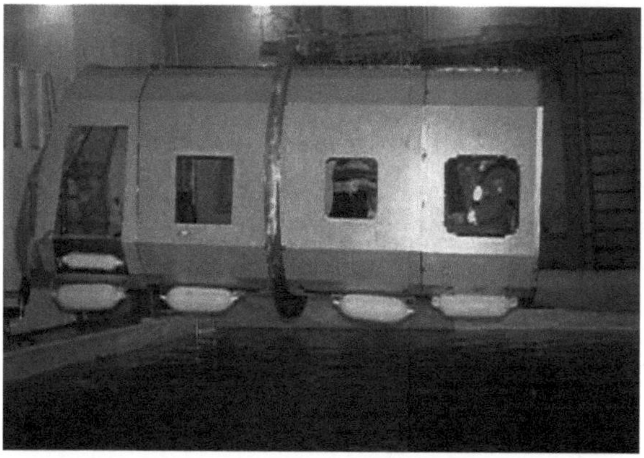

Figure 2.4 Modular Egress Training Simulator (METS™) based on the CH124 airframe (1987). *(From Taber and Bohemier (2014) with permission.)*

rigid-hull inflatable boats. Although there are a number of other organizations that design and build UESs (Chapter 7), the METS™ is currently considered to be the world standard. For example, as of this writing, more than 100 models of the METS™ are in use all over the world.

2.1.2 Development and Implementation of HUET

After considering the history of simulator types used for HUET, it is important to focus on what is actually occurring for the individuals involved in a real-world helicopter ditching event. The following vignette describes a hypothetical scenario that outlines the likely experience from the perspective of someone who has just ditched. The vignette also presents information that considers the situation if there are other individuals in the cabin during the event.

Let's imagine that you are an oceanographic technician flying over Lake Superior onboard a small twin-engine helicopter. You, along with a research scientist and two pilots, are returning to your vessel following a series of stationary-flight water samples. The samples are collected by bringing the helicopter into a low hover and then lowering a probe into the water. The data collected from the samples will help to track water temperature and salinity across the Great Lakes. You and the research scientist have finished labeling your samples and are settling in for the return flight to your vessel. You have just sat back in your seat and secured your four-point restraint harness for the flight back. One of the pilots informs you, via the aircraft's internal communication system, that they will be climbing to 3000 feet and that your estimated time of arrival back onboard will be approximately 40 minutes. You loosen your seatbelt and remove your headset, stretch out in the seat, close your eyes and quickly drift off to sleep.

You wake up when your colleague slaps your leg and starts yelling something indecipherable. The cabin is beginning to fill with smoke and you can feel the aircraft's rapid descent. Although you cannot see a fire, there obviously is one somewhere onboard the helicopter. You can faintly make out the sound of an alarm, and the instrument panels in cockpit are alive with flashing lights. The research scientist is scrambling to strap down the water samples and scientific equipment scattered haphazardly around the rear of the helicopter. You quickly place your headset back on your ears and hear the pilots reciting their emergency procedures checklist for an engine fire. They are methodical completing all of the steps on the checklist, but their voices sound excited and tense. They activate the engine's extinguishers and shut down both engines. You hear the copilot say it might be an electrical fault. The pilot is initiating a Mayday and relaying your position. You can't tell if anyone has received the message or not. Things are happening very quickly. Your adrenaline is pumping. Your mind is racing.

The copilot quickly informs you that the helicopter is in an emergency descent headed to the surface and to prepare for ditching. You immediately push back in your seat and start to tighten up your seatbelt. Your hands are shaking and you can barely grab and pull on the ends of your lap belt. The research scientist is sitting across from you and is frantically trying to locate his harness and buckle up. You glance out the window and notice with an almost surreal indifference that the surface of the water is coming up to meet the helicopter with surprising speed. The pilot shouts into the radio "Brace, brace, brace for impact!" and you feel a sickening sensation in your stomach as the aircraft decelerates and the nose of the helicopter pitches skyward. As you manage to get your arms crossed and grab your shoulder harnesses you can see that the research scientist is just clicking in his lap belt. His face is a textbook image of shock and awe. You drop your chin down to your wrists and shut your eyes tight.

The moment of impact is a violent and intense event. The g-force of the helicopter's fuselage striking the surface of the water drives the air from your lungs as you feel the world turning sideways. Although your eyes are closed your brain is simultaneously registering movement, noise, pressure, confusion, and uncertainty all at once. The cabin is rapidly filling with icy cold water, which seems to slap you in the side of the face and then, in a heartbeat, you are completely submerged upside down. Your initial instinct is to get away: get away from the helicopter, get away from the cold, and get away from the water. You open your eyes to try to see where you are and what's happening but the 9 °C water temperature makes your eyes ache. What little you can see is blurry and, despite the fact that it is a bright clear afternoon, there is very little light inside the helicopter's cabin.

It has been only 8–10 seconds since you crashed into the water; however, the desire to breathe is quickly becoming overwhelming and you feel as if you could chew through the side the aircraft with your teeth if it meant you could just get to the surface and take a breath of air. You remember that you are wearing a compressed-air emergency breathing system (EBS), which is in a pocket attached to your aviation vest. Your right hand runs down your body along your rib cage and you feel the distinctive shape of the second-stage regulator. You pull it free of the dust cover and awkwardly insert it into your mouth and take a deep breath. You are expecting the sweet relief of air but the regulator delivers a mouthful of icy cold water instead. You begin to gag and choke as you feel panic wickedly trying to take control of your actions. You press the purge assist button on the face of the second-stage regulator as you blow water out of your mouth and then pull back another desperate breath. There is still some water present but this time you are rewarded with a breath of air. With some effort, you forcefully blow the residual water from your mouth with the second exhalation, and then pull back another more confident lungful of air.

You are breathing! Clarity of thought is returning. You can literally feel your heartbeat slowing in your chest. You can breathe, so you can think. You know you need to escape. You open your eyes again, only for a second, but can still see virtually nothing—only shadows and darkness—the cold forces your eyes closed once more. You take another breath and then run your hand along your leg to your knee and from there to the bulkhead. Your arm slides into the depression of the push-out window you are sitting beside. You push but nothing happens. You quickly readjust your body position and, using your elbow, lean with all your strength into the corner of the window. It gives way under your force. You experience a brief moment of excitement and disbelief as you realize a path to safety has just opened and that you may yet survive this nightmare.

Your outboard hand locks firmly onto the bottom of the open window frame. The window frame is the way out and the hand maintaining that reference point is your lifeline to freedom. Using the opposite hand you search for and locate the rotary release mechanism on your restraint harness. It is hard to hold onto. Your hands are almost numb with cold. You feel the mechanism twist and something gives and you pull yourself hard toward the window frame. On the way, your body comes to a sudden halt. Initially you think something has snagged you somewhere on your helicopter immersion suit, so you pull harder but realize you are stuck fast to your seat. You are still in your restraint harness. It failed to fully release. You take another breath and try again to operate the release mechanism with your cold, weak hands. Again you feel the mechanism click and now you are moving, pulling yourself through the window frame, not really sure which way is up or which way is down. When your body is half inside and half outside of the helicopter you place a hand on either side of the window and push hard. As you clear away from the submerged and inverted helicopter you take another breath and start clawing toward the surface. After what feels like an eternity your face breaks the surface of the water, you keep the second-stage regulator of your EBS in your mouth until you're sure you can breathe, and you remember you should inflate your aviation vest to conserve energy and make yourself more visible. You struggle to locate the inflation toggle but your hands are so cold and numb that the task seems virtually impossible. You lean onto your back in the water and are able to see the inflation toggle. Your hand is freezing cold but you manage to get the vest inflated and start to scan the water for other survivors. Both of the pilots surface seconds after you, so you begin calling out to them and swimming in their direction. After the three of you have regrouped and the bewilderment of what has just occurred passes, you realize that the research scientist is still inside the aircraft. After 2 or 3 minutes have passed the reality that he has failed to escape becomes sickeningly clear.

The vignette illustrates how quickly most helicopter ditchings take place (roughly less than 15 s of warning) (see Brooks, MacDonald, Donati, & Taber, 2008 for further details). Fifteen seconds or less leaves little time to process what is happening and to figure out a plan of action needed to survive the impact. Once the impact is over, you still need to deal with the shock of sudden cold-water immersion (assuming you are in a cold-water environment—see Chapter 4), overcome the disorientation and confusion associated with the rapidly flooding and inverting helicopter, and recall and implement the procedure to escape (Leach, 2004; Mills & Muir, 1999). Every situation will be different. The variables and complications are almost endless, and the psychological factors will also be different for each individual

(Chapters 3, 5, and 6). By understanding that there is no cookie-cutter example of how a ditching event may occur, we can become better at training passengers and aircrew to deal with a myriad of possible scenarios and, in doing so, increase the likelihood that most, if not all, will survive.

Despite the obvious goal of HUET programs to increase survivability, global HUET program development and implementation are almost as varied as the circumstances surrounding helicopter ditchings. The methods and techniques used in training extend across a wide spectrum that ranges from intense high-fidelity exercises using aircraft-specific simulators and procedures conducted in highly realistic sea states with storm conditions all the way down to a plastic chair bolted to a PVC frame that is manually flipped upside down in a local swimming pool (Taber, 2010, p. 211–290; 2014; Wells, 2010; see Chapter 7). The reasons for the varying degrees in training fidelity include cost, time, equipment availability, government and industry standards in a given geographic area, the attitude of the trainee, and the passion, skill, and knowledge of the instructional team delivering that program.

A variety of HUET methodologies are delivered by educational institutes, private companies (both large and small), civilian aviation organizations, and military groups around the world. As stated above, all of the programs have the same basic common denominator: to save lives. However, some are better than others and the hierarchy is not necessarily determined by the type of organization or company delivering the training.

2.1.3 Types of Programs

2.1.3.1 Military Programs

Elite military HUET programs such as those delivered to the American or Canadian military are generally considered to be at the top of the training food chain. These programs are tailored to a specific aircraft type. The theoretical and practical course curriculum targets specific aspects of the event for each member of the aircrew. To enhance this specificity in the performance of critical tasks during an emergency, aircrews are encouraged to train together whenever possible. Based on accident investigation report findings, the program addresses the aircraft type, flying conditions, and common mission profiles during flight with an emphasis on critical phases of flight (e.g., takeoff, hover, landing, and approach). Night operations also receive extra attention because of the reduced visibility leading up to and during the ditching and the increased difficulties associated with performing postegress survival skills, and because rescue operations become more difficult at night (Taber, 2010, p. 211–290, 2014).

When training military personnel, HUET simulators are typically configured to replicate the operational aircraft type (Figure 2.5). Emergency exit locations are aligned as precisely as possible with regard to seating and equipment locations. Props, such as helmet bags, stretchers, life rafts, workstations, flying controls, overhead or center consoles, sensor operator stations, helicopter emergency egress lighting systems, internal egress guide bars, and other relevant egress landmarks are incorporated into the simulator to ensure that personnel develop an egress process in training that can be transferred directly to the real aircraft (Chapter 7). To further enhance the training environment, the aircrew wear representative flight suits and aviation vests, aircrew harnesses, survival backpacks or seat-packs, and flight helmets, which may be fitted with mock night vision goggles and communications cords. In the case of helicopter-borne troops or special operations personnel, equipment may include individual restraint harnesses, body armor, gas masks, rubber weapons, fast ropes, or rappel harnesses. Search and rescue personnel or flight engineers may also train using monkey tails or hoisting points.

Trainees are taught to egress from primary, secondary, and tertiary crew positions. Exits may be in place, already jettisoned by instructors or other trainees, or set in the open position such as sliding doors or aircraft ramps.

Figure 2.5 Underwater egress training simulator interior military configuration example.

Trainees may be seated or kneeling, restrained or free standing. They may be sitting on the floor with their feet hanging from the sill, connected to ropes or rappel anchor points. Dependent on the aircraft type and personal equipment worn, trainees may perform egress techniques from crash-attenuating seats and they may be expected to disconnect communications cords or remove their flight helmets prior to egressing. Often crew will be required to wait for another crewperson to egress before it is their turn to move. Although this is a realistic situation for many aircrew, it is not currently an egress training sequence that is fully incorporated for offshore passengers.

As training exercises or evolutions progress and become more intense, wind, waves, rain, low-level light or complete darkness, and a range of sound effects will be added to increase the feeling of realism. All or some of this training may also involve the use of compressed-air EBSs (Civil Aviation Authority, 2013; see also Chapter 8). Following or prior to their inverted training exercises trainees will also perform surface or hover evacuations as a group. They will be required to inflate aviation vests, board a life raft, secure the raft canopy, and perform sea survival skills in a simulated storm until rescued.

This type of training complexity and integrated task requirements may appear complex at first. However, as the instructional staff become more familiar with the aircraft layout and configurations including the type of equipment being used by each trainee (including their roles and duties in the helicopter), it becomes easier to identify the specific interactions necessary to prepare individuals for a possible ditching. In the multifaceted and dynamic world of military aircraft ditching programs the key to delivering successful training is understanding the specific needs associated with a particular organization or unit.

2.1.3.2 Paramilitary/Civilian Programs

Civilian aviation programs are also quite diverse. They may include contract search and rescue (SAR), law enforcement, aerial surveillance, air ambulances, and offshore line pilots to name a few. Similar to the military training, the key to developing and delivering applicable programs is to take the time to research and understand the organization. Given that civilian aviation groups do not typically operate at as high an operational tempo as the military, their training intensity is normally lower. Also, civilian aviation communities may or may not be mandated to participate in ditching training. Others, such as offshore line pilots who spend the majority of their flying time transporting oil and gas workers to and from offshore installations or vessels, are more likely to be required to take HUET as part of a service contract. Regardless of the legislative requirements, flight personnel who fly over water typically seek out a ditching program whether they are required to or not. Although the training may not require the same level of intensity as military programs, training providers must take the time to research the organization and develop training programs that are applicable to the target group. Once these programs have been developed and are put into practice, it is important to stay current and update them as required.

When a training provider develops a generic helicopter or fixed-wing ditching program that does not take into consideration differences between the unique settings/equipment and delivers that one program across both platforms, it is the end user that suffers. For example, if a training provider were delivering a one-day HUET program and included oil and gas passengers, airborne surveillance personnel, pilots, and air medics in the same class it would be impossible to address all of the specific requirements for each group. A theory portion of the training might address some commonalities associated with basic helicopter emergencies over water and actions to be taken after impact; however, without the specificity of operational differences, there is a risk that the trainees would find the information to be too basic to be of any value. The practical training exercises for such a diverse group would likewise require a very rudimentary exposure to helicopter underwater egress and would be unlikely to properly prepare the trainee for a real-world event. Generally speaking, jurisdictional HUET requirements greatly influence the level of realism used for the UES and training protocol. Without a clear understanding of the specific issues, it is not unreasonable to assume that an operator relies on regulatory requirements and assumes that the training provider is doing the best job possible to prepare his or her trainees. Unfortunately, from a training perspective, "you don't always know what you don't know."

2.1.3.3 Offshore Specific Programs

Standardized training programs work well when delivering the program to large groups of people who are flying in the same aircraft types and under the same conditions all the time. The best example of a standardized HUET program would be that delivered to oil and gas workers. For example, in the North Sea in 2013, according to the UKCS Offshore Workforce Demographics Report, a total of 61,892 people traveled offshore (UK Continental Shelf Offshore Workforce Demographics Report 2014;

Oil & Gas UK, 2014). The number of core workers (those who spent more than 100 nights offshore) was 27,749. Globally there are in excess of 100,000 people flying offshore each year (UK Continental Shelf Offshore Workforce Demographics Report 2014; Oil & Gas UK, 2014)

When dealing with these large numbers of people, the approach should be one of quality, consistency, and safety. Groups such as OPITO (formerly known as Offshore Petroleum Industry Training Organization), OLF (Oljearbeidernes Fellessammenslutning; Norwegian Oil Industry Association), or CAPP (Canadian Association of Petroleum Producers) all strive to achieve the same results; however, their approaches to training development and delivery differ considerably.

Canadian HUET

For example, CAPP (2013) Atlantic Canada Offshore Petroleum Industry Standard Practice for the Training & Qualifications of Personnel requires HUET for everyone flying to the offshore area. CAPP HUET competencies reflect a clear understanding of what skills individuals would be likely to need in an actual ditching. The specific task requirements were developed in consultation with HUET program instructors and based on empirical research findings. The skills are a component of the initial entry and refresher basic survival training (BST and BST-R), which all workers in the Canadian sector must have prior to traveling offshore by helicopter. Visitors (and individuals who will not spend more than 7 days offshore during a 12-month period) must also complete an Offshore Survival Introduction (OSI), which includes the same HUET competencies as those required for the more permanent offshore workforce.

CAPP requires workers to complete HUET while using representative emergency exits and push-out escape windows (Figure 2.6). "Representative" in this case means close replicas of the types of exits used in the helicopters transporting workers offshore. Given that passengers may be seated on either the port or the starboard side of the aircraft, CAPP also requires personnel to perform an inverted egress from both sides of the simulator. Trainees must also demonstrate that they can escape from a seat in the fully stroked position while in a simulated crash-attenuating seat (Taber, 2013). Understanding that not every passenger will be afforded the opportunity to sit immediately adjacent to an exit, trainees are also required to demonstrate an inverted underwater egress from a seat not located immediately next to an already open exit.

Figure 2.6 High-fidelity simulation of an S92 cabin interior for offshore HUET program.

The CAPP HUET progression includes a minimum of five runs including the surface abandonment exercise. The surface abandonment exercise is initiated from the seated and strapped-in (four-point rotary harness) position. When directed by the instructor the trainees must don their hoods and zip up their suits. Trainees then brace for impact, descend to the surface of the water, and prepare for the abandonment. The Canadian sector abandonment exercise, which is not commonly taught in other regions, asks the trainees to decide if it is safe to jettison the exits or not when the UES comes to rest on the surface of the pool. This decision is based on an assessment of the situation (engines have powered down and if the sea conditions are favorable—waves are not breaking over the window—and there is no obvious threat such as fuel or fire outside). Typically, HUET programs advise trainees to wait to jettison exits if the aircraft is upright and stable on the water until directed to do so by the aircrew. Trainees in the Canadian sector are taught that once the exit has been jettisoned, they should establish a reference point and wait until it is their turn to move to the life raft. The reasoning for this focus on self-evaluation of the situation is that the aircrew may be incapacitated in the ditching, may not be able to use the internal communications, or may neglect to instruct the passengers to perform this essential step owing to the complexities of the unfolding situation. Having the exit open and a hand secured on the window frame provides the passenger with a definitive advantage if the aircraft inverts prior to or during the abandonment. Trainees then release their harnesses and, while maintaining three points of contact in the aircraft, move to and board an aviation life raft. At this point in the exercise trainees are instructed on the techniques to secure the raft canopy, inventory their survival equipment, treat the injured, and prepare for rescue.

The CAPP standard requires a minimum of four inverted HUET runs. Unlike OPITO or the Norwegian OLF, CAPP permits training providers to include additional inverted runs into their progression sequence if they desire to do so as long as it adds quality to the program. During run 1, the UES stops at the surface of the water. This allows the trainee to jettison his/her exit and establish a reference point. After inversion, the trainee releases his/her harness and egresses to the surface. During run 2, the trainee is directed to maintain his/her brace position throughout the rollover, wait for all motion to stop, locate and jettison the exit, establish a reference point, release his/her harness, and then egress to the surface. Run 3 is essentially the same as run 2 but must be completed using the opposite hand. This is accomplished by either moving to the opposite side of the UES or switching from a forward-facing to a rear-facing seat position. Either run 2 or run 3 must be completed by the trainee from a fully stroked seat position. Run 4 requires the trainee to move to an already open exit and egress to the surface. This run teaches the trainee how to move from one position to another in the event he/she had to follow another passenger out or if he/she were required to egress through a secondary exit.

Trainees are taught to recognize and respond to emergency announcements from the aircrew and how to identify other indications of danger during flight (e.g., fire or mechanical problems). Because operators in the Canadian sector conduct flight operations over cold water (i.e., below 15 °C), this training must be completed while wearing a representative helicopter passenger transportation suit, also known as an immersion suit (Chapter 9). During their HUET exercises trainees are required to inflate their suit, deploy their spray shield, put on their gloves, and, in some cases, simulate the activation of a personal locator beacon. Beginning in 2009, CAPP also required that all workers complete compressed-air EBS training in conjunction with HUET training in both a dry and a wet environment (Chapter 8). This includes practice deploying, purging, and establishing a breathing pattern while submerged and inverted in the water.

Norwegian HUET

The Norwegian OLF requires all workers to participate in the Basic Safety and Emergency Preparedness Course (GSK) (Norwegian Oil & Gas Recommended Guidelines for Safety and Emergency Preparedness Training, translated version, version 21, 2013). The HUET components of the Norwegian program comprise aspects similar to those outlined in the CAPP guidelines. The GSK uses a "breathing lung" instead of a compressed-air EBS; however, it is also integrated into both wet and dry training requirements. Following a training session with the breathing lung in the water in which trainees must demonstrate their ability to breathe while submerged, GSK trainees undergo six HUET runs using a HUET simulator while seated adjacent to an exit.

OPITO HUET

OPITO is undoubtedly the most recognized offshore HUET program in the world and they require that trainees take part in HUET training for both the BOSIET (Basic Offshore Safety Induction and Emergency Training) and FOET (Further Offshore Emergency Training). OPITO (2013) also has a standalone HUET program. In addition to the BOSIET, FOET, and HUET programs, which are designed for passengers transiting over cold water while using an Air Pocket EBS rebreather, OPITO also has tropical variants on these programs (T-BOSIET, T-FOET, and T-HUET) for warm-water environments, which do not require the use of an EBS or an immersion suit. At the time of writing, OPITO is implementing the use of a compressed-air EBS unit but training is being provided only in a dry environment and it is not being employed in the HUET simulator (Chapter 8).

With the exception of the tropical variants, all OPITO training is conducted using the Air Pocket rebreather EBS unit, which, like the Norwegian breathing lung, allows the user to rebreathe his/her own air to aid in evacuation or underwater egress. All trainees must demonstrate proficiency in the use of the EBS prior to using it in the HUET simulator.

The OPITO HUET programs all include a surface abandonment exercise following a controlled landing into an aviation life raft. These programs require trainees to board the raft directly from the simulator, as well as entering the raft from the water. Trainees then carry out initial actions in the raft including cutting mooring lines, raising the raft canopy, and checking the raft for damage. Secondary actions such as posting lookouts, activating electronic signaling devices, and performing first aid are demonstrated and explained. Trainees also wear representative immersion suits, aviation vests, and EBS units during all practical HUET exercises. During tropical training trainees wear aviation vests and EBS units but are not required to don immersion suits.

The HUET run progressions for the BOSIET and HUET programs are the same. The first three runs are conducted with the trainee seated adjacent to an exit and with the simulator entering the water upright and then sinking straight in. Run 1 does not require the trainees to use their EBS or to jettison their exit. Run 2 requires the trainees to establish and activate their EBS prior to submersion but does not require them to jettison their exit. Run 3 requires the trainees to use their EBS and to jettison their exit while submerged. The final three runs (4, 5, and 6) are all 180° inverted training runs. Run 4 does not require the trainees to use their EBS or jettison their exits. Run 5 incorporates the use of the EBS by having trainees establish and activate the units prior to submersion without the requirement to jettison their exits underwater. Run 6 requires the trainees to use their EBS and to jettison their exit while submerged.

The FOET program, which is the BOSIET refresher training, requires only three HUET runs. Run 1 is conducted with the trainees seated adjacent to an exit and with the simulator entering the water upright. The trainees establish and activate their EBS but are not required to jettison their exit. Run 2, which is also conducted with the simulator entering the water upright, requires the use of the EBS and requires the trainees to jettison their exit while underwater. During run 3 the trainees establishes and activate their EBS and jettison the exit while submerged from a fully inverted simulator.

Because the EBS units require time to establish and activate prior to being ready for use, all OPITO HUET egress exercises that include the use of EBS necessitate that the simulator be stopped at the surface of the water prior to submersion. Once the trainees are breathing from their EBS units, the simulator is either submerged upright or inverted. This is also true of the Norwegian GSK program.

The tropical BOSIET, HUET, and FOET programs require trainees to perform four HUET exercises. Runs 1 and 2 are conducted upright and runs 3 and 4 are conducted inverted. All exercises are performed with the trainees holding their breath and only runs 2 and 4 require the trainees to jettison an exit underwater.

2.1.3.4 HUET Program Summary

In Canada, the required underwater HUET exercises are the same regardless of the offshore course being completed (e.g., BST, BST-R, and the OSI). Trainees are required to perform a minimum of four inverted HUET runs with the first run being conducted after the exit has been jettisoned at the surface. Each subsequent run occurs with no stops at the surface and is intended to simulate no-warning water entries. Trainees must demonstrate that they can egress the aircraft using either hand. They must also demonstrate that they are capable of traveling from their seat to another exit not immediately adjacent to them.

The GSK HUET exercises require trainees to escape from a submerged and inverted simulator in five of the six training runs. A primary difference between the HUET requirements for the GSK and CAPP programs is the number of times the UES stops at the surface prior to sinking and inverting. For the GSK program only the final run simulates a no-warning water entry without a stop at the surface. The reasoning for this is to allow trainees time to prepare their breathing lungs. Five of the six runs require the trainees to use their breathing lungs while performing the exercise. The HUET exercises are identical for the GSK refresher training.

In contrast to the Canadian version of HUET, all of the OPITO egress exercises are conducted with the trainees seated immediately adjacent to a push-out exit. No mechanical exits are used during underwater escape training although they may be used during surface abandonment exercises during evacuation to a raft. Trainees are also not required to move to an alternate or secondary exit point nor are they required to switch sides (port or starboard) during their training.

2.1.4 Vital Information

One of the primary goals of HUET is to ensure that individuals are as prepared as possible to survive an aircraft ditching. Additionally, it is important for the trainees to be confident that they can accomplish the egress tasks. This is true whether training oil and gas workers or Special Forces groups. No matter the audience, a critical piece of information that should be conveyed to all trainees is that helicopter and fixed-wing ditchings are very survivable (typically around 76 and 90%, respectively) (Civil Aviation Authority, 2014; Brooks, 2007; Brooks et al., 2008; Taber & McCabe, 2006, 2007; Taber, 2010, p. 211–290, 2013, 2014). This is an important message in training because if trainees enter a program believing that they will not survive a ditching, they are less likely to pay attention to the information or skill sets being provided.

2.1.5 Optimal HUET Program Development

Determining what information should be included in a HUET program should be carefully considered. It should include information from actual ditchings/water impacts as well as human performance research. By studying accident reports, listening to survivor testimonies, watching videos of aircraft accidents, reading papers relevant to the subject, and continuously engaging the scientific community for answers about the human factors (performance under stressful situations) involved in underwater egress, it will be possible to create a program that prepares trainees to respond appropriately during life-threatening emergencies.

The integration of this information allows real-world helicopter or fixed-wing ditching events to be organized into four phases. The preimpact, impact, postimpact, and survival/rescue phases represent specific points in each event and should be considered in isolation as well as integrated components of the entire survival situation. For example, it is important to recognize that each phase requires a slightly different skill set.

2.1.5.1 Preimpact Phase

This is the point at which an individual realizes he/she is in danger. This may occur with or without altitude and, thus, with or without warning (Brooks et al., 2008; Taber, 2014). Regardless of the time available, it should prompt an emergency response from the aircrew and/or passengers (see Chapter 3). Trainees need to understand the indications of danger (e.g., an announcement from the cockpit, loud bang, smoke, rapid descent). However, it is important to maintain a good level of situational awareness—particularly during the critical phases of flight. For pilots, a warning light or an alarm on an instrument panel, unusual vibrations, engine shutdown in flight, bird or lightning strikes, or a loss of engine or transmission fluids all represent situations that require some form of a response.

Preparation, depending on the individual requirements (e.g., aircrew versus passenger), may involve securing personal and/or loose objects in close proximity, ensuring the immersion suit is watertight, donning hoods, securing or tightening the seat belt, bracing for impact, and mentally preparing for whatever may be coming next. If the aircraft is at a low altitude and an emergency develops, there will be considerably less time to prepare (Brooks et al., 2008). At lower altitudes, the emphasis should be on bracing to increase the likelihood of surviving the impact.

A good brace position will reduce injuries and unconsciousness during impact, the possibility of being hit by flying debris, overall body profile (surface area) to inrushing water if maintained during a rollover, and disorientation if the aircraft does invert (Cheung, Hofer, Brooks, & Gibbs, 2000). By bracing and reducing injuries, individuals will be better able to perform vital actions for survival (e.g., jettisoning exits, inflating vests, boarding rafts, activating signaling devices, tying knots, firing flares, or assisting injured survivors).

2.1.5.2 Impact Phase

With the exception of a controlled landing on the surface of the water, the impact phase is something that cannot realistically or safely be re-created using a HUET simulator (see Chapter 7). However, impact forces are a significant survival factor in any ditching. If the forces are great enough, they may result in damage to the airframe, jammed exits, shifting equipment, a multitude of injuries, and possibly unconsciousness. The impact forces may also cause a release of aviation fuel on the surface of the water, which will further complicate the situation during abandonment or egress from the aircraft.

2.1.5.3 Postimpact Phase

This phase of the event will result in the helicopter being either upright on the surface or inverted (at the surface or sinking). If the helicopter remains at the surface, it is possible that the aircrew will elect to perform a surface evacuation (either dry or wet abandonment into a raft—see Brooks & Potter, 1998). Upright on the surface with emergency flotation deployed is the best situation and has the highest survival rate (Taber & McCabe, 2006, 2007). The second possible position is inverted (partially or completely 180°) but still afloat at the surface (Coleshaw & Howson, 1999). In this case, individuals will need to make an underwater escape but will not have far to travel to reach the surface (Taber & McGarr, 2013). Taber and McCabe (2007) clearly showed that next to floating upright, survival rates were significantly higher if the helicopter remained at the surface. The third possible position is inverted and sinking. In this case, following egress, individuals will need to reach the surface before breaking their breath-hold or before they run out of air if equipped with an EBS (Cheung, D'Eon, & Brooks, 2001; Taber & McCabe, 2009).

2.1.5.4 *Survival/Rescue Phase*

This phase of the events begins from the moment of egress until safely back home (Taber, 2014). Once clear of the aircraft, individuals may find themselves alone or with other survivors, with or without a raft; it may be daytime or nighttime, and the seas may be calm or stormy (Knass, 2013). Rescue may come via a vessel or aircraft, may be military or civilian, or may be any vessel of opportunity. Whenever possible getting out of the water is the best option, especially if the water is cold (Chapter 4). This can be achieved using life rafts if they are available or by climbing onto the belly of an inverted helicopter if it is safe to do so.

If in a raft, individuals should stay with the aircraft unless there is a threat that debris or sharp edges on the helicopter will damage the raft or potentially cause it to roll over. If the sea state is unfavorable and individuals are in a raft, sea-sickness medication should be taken as soon as practical. Electronic signaling devices should be activated and pyrotechnic signaling devices should be prepared for use. Lookouts should be posted to search for other survivors or rescue opportunities and a sea anchor should be set to reduce drifting. A plan of action should be discussed to ensure that all survivors know what is expected of them when rescue does arrive.

By describing the four phases of a ditching, trainees will be provided with information that encompasses the entire ditching event (i.e., from onset of the emergency to rescue). This framework can be used to develop theory presentations and provide a realistic structure for practical training exercises. Despite being the same for all ditching events, the details of this plan will be slightly different dependent on the tasks and the individuals carrying them out.

2.1.6 Practical HUET Safety Protocols

Preparation is essential when delivering any safety training-based program. Practical exercises must, above all else, have the safety of the trainee and the instructional staff as the top priority. A program can be developed and delivered with high-fidelity aircraft ditching exercises that are intensely realistic (Chapter 7). However, if it is not possible to deliver the training in a consistently safe environment, then the end product is not dependable or repeatable and is unlikely to properly prepare the individual to transfer the information to the real world.

Risk assessments must be conducted to ensure that applicable mitigation strategies can be incorporated as required. Training organizations should select the most appropriate risk assessment technique that works best for their particular situation and then apply that model during the development of all practical exercises (Glendon, Clarke, & McKenna, 2006). If new risks or hazards are identified later, then the risk assessments should be updated accordingly.

Once made aware of the risks and mitigating factors, members of the HUET instructional team should focus on the communications and procedures required to deliver the training. A variety of communication and procedures can be used by the training organizations to ensure that safety is the primary focus. Identifying the roles and responsibilities for each response team member is critical to ensure that the program is operated safely. Typically, there are three core roles within the HUET instructional team: HUET instructors, HUET divers, and hoist operator. It should be noted that some organizations do not use divers. Instead, they use additional instructors to maintain staff-to-trainee ratios. Although each person has a part to play, all members of the instructional team are responsible for trainee and staff safety at all times.

2.1.6.1 Operational HUET Process

Normally the lead HUET instructor is in control of the exercise with additional instructors being brought in as required. Trainees rely on the experience and guidance of the instructors to bring them safely through what some may consider stressful training (Coleshaw, 2006; Robinson, Sünram-Lea, Leach, & Owen-Lynch, 2008) and to provide them with the required skill sets and learning outcomes upon completion of the program (see Chapters 6 and 8). Instructors are responsible for briefing and debriefing trainees during exercises, answering questions regarding procedures and techniques, responding to problems inside the egress simulator, evaluating performance, and keeping the lesson on schedule.

When a trainee enters the UES for the first time, the instructor should direct him/her to an assigned position for the upcoming exercise evolution and assist with the seat belt, if required. It is particularly important to ensure that first-time trainees are directed throughout the entire sequence of events. There should be no time at which individuals are unsure of what is expected of them while inside the UES. New trainees will often have a myriad of questions as soon as they enter the simulator. "How do I open this window" or "how does this seat belt open" or "when should I employ my EBS" are typical questions for those who have not previously completed HUET. The instructor should reassure the trainees that all of their questions and concerns will be addressed before the start of the exercise. However, addressing the specific queries should not occur until the trainee is fully strapped into the seat and the instructor is satisfied that the trainee is safe to proceed. If, for example, a new trainee is fixated on the correct technique to jettison an exit and is also trying to figure out how to properly connect his/her seat belt while the instructor is explaining how to adopt a proper brace position, the probability of vital information being missed or forgotten is virtually guaranteed. Successful egress from a submerged and inverted helicopter requires that individuals follow a set procedure and training should also follow a set procedure.

Instructors should start the pre-simulator boarding process by explaining that an exercise briefing will be provided as soon as the trainee is safely strapped in and ready to go. Once the trainee is secured in his/her seat, the instructor should conduct a visual "head to toe" assessment on the individual. The instructor should be looking for snag hazards, improperly stowed equipment, immersion suits that are not fully zipped up, proper routing of EBS regulators, and seat belts that are not properly secured with release mechanisms accessible, and the instructor should note the position of the seat emergency release system (ERS). The ERS (Figure 2.7), when activated, will completely release the seat belt in the event of an emergency,

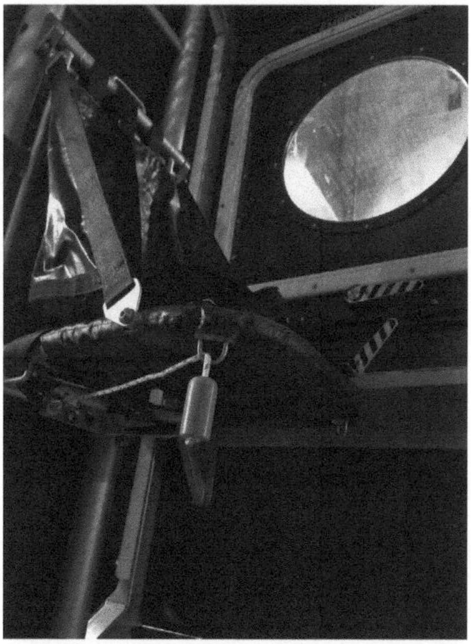

Figure 2.7 Emergency release system used to disconnect the entire seat belt if necessary. (Note: the handle on the end of the rope is connected to a quick-disconnect device that holds the end of each seat belt strap.)

thus allowing for a fast response in the event of a panicked or distressed trainee.

The instructor should also be planning how he/she will assist the trainee out of the simulator and to the surface should the HUET run go poorly. Once all of this has been determined, then the instructor can deliver a short concise briefing knowing that the trainee is now paying attention to the described procedure and not focusing on other concerns.

When all trainees are ready to proceed, the instructor should verify the readiness of any other instructors inside the simulator and then initiate the call to the hoist operator and divers to proceed with the HUET run. This may include ditching to the surface (the simulator will stop at the surface of the water) or an upright or inverted HUET submersion. Regardless of the nature of the upcoming run, the instructor should always position himself/herself within reaching distance of the trainee so that he/she is able to assist if required.

Nonemergency HUET Response

If the exercise involves the simulator being submerged, the instructor should maintain his/her orientation to the simulator. In other words, if the simulator inverts, the instructor should also invert with the simulator. By remaining in the same orientation as the simulator, the instructor can provide a more effective response to trainee difficulties or emergencies. When responding to trainee problems underwater, the instructor should follow the 3R rule of problem management (early Recognition, fast Reaction, and appropriate Response). Different situations require different applications of the rules. For example, if a trainee is calm but is having trouble locating a mechanism to jettison an exit, the instructor should quickly recognize the issue. Following an unsuccessful attempt, the instructor should be prepared to place the trainee's hand in a position to activate the mechanism before the situation becomes problematic. The instructor's response should then be to allow the trainee to complete the run and emphasize the proper technique to locate the exit mechanism during the debriefing.

Emergency HUET Response

In a case in which, immediately following the UES inversion, the trainee begins to panic and struggle, the instructor should recognize that the individual has lost control and requires immediate assistance. The appropriate response to this emergency would be to: (1) jettison the closest exit if not already done, (2) locate and activate the ERS while maintaining contact with the trainee (release seat belt at the primary mechanism only if safe to do so), (3) quickly guide the trainee out of the simulator, and (4) ensure he/she has safely made it to the surface. If it is not possible to guide a trainee out of the simulator, the instructor should indicate, through appropriate and recognized hand signals, that there is a need for an emergency raise. The hand signal should be recognized by the HUET diver, who should initiate the emergency raise by calling out to the hoist operate, "Raise." As the simulator is raised slightly out of the water, the instructor should direct the trainee to an air gap and ensure that the person is breathing. Once all trainees have egressed the simulator the instructor should check that any other instructors are also safe. Having completed the exercise, the instructors should leave the simulator in a controlled manner, debrief their trainees, and prepare for the next training evolution.

HUET divers, when employed, play a critical role in the team. They act as the eyes of the hoist operator by constantly monitoring what is happening inside the simulator while it is submerged and then communicating when staff or trainees are in distress by initiating an emergency raise. They also act as a safety backup for instructors by visually verifying that everyone is safe at the end of each HUET run (i.e., not injured, entangled, or trapped inside the simulator). In addition to recovering exits off of the bottom of the pool and resetting them following each HUET run, divers are responsible for a myriad of additional duties during sea survival training (e.g., maneuvering life rafts, providing direct supervision of trainees in the water, and assisting with exercise logistics). Like all members of the HUET team, the divers' primary responsibility is the safety of the trainees. They are required to exercise good judgment and to make quick decisions about when to intervene or assist if a trainee is experiencing difficulty or an emergency raise is required.

There are a number of specialized operational hazards that the divers face. For example, the divers may be struck by prematurely jettisoned exits if they are not in a safe area (dictated by the design and physical layout of the pool area). Ideally they will be positioned in the shallow end where they can stand and wait until they are required to perform their duties. However, some training pools do not have a shallow end so the divers must remain afloat at the surface until ready to descend. There is also a possibility of becoming snagged if positioned too close to the UES while it is rolling. Finally, they may be required to control and assist panicked trainees on the surface. Regardless of the facility layout, the "safe area" should permit the divers to remain in an area where they will not be accidently struck by a falling exit or positioned underneath the simulator as it is descending to the water.

Once the diver is submerged, he/she should be looking toward the surface in anticipation of the simulator entering the water. Depending on the exercise, the simulator may be entering the water upright or may be inverting. The divers should quickly move to a position where they can be looking into the simulator as soon as it has stopped moving. If they are too fast they may be struck by the simulator while it is in motion. If they are too slow they may miss a struggling or panicked trainee who requires immediate assistance. The diver should ensure that upon egress, the trainee is safe and moving in a controlled manner on the surface. After they have confirmed the trainee's safety, they should then check to ensure that instructors are safely leaving the simulator. Once the safety of all personnel is confirmed with an OK signal, the divers can descend to collect the exits and reset them for the upcoming exercise evolution.

The rules and legislative requirements for HUET divers vary significantly from region to region. HUET divers should undergo a training program that teaches them how to work and respond in this highly specialized environment. Training prerequisites will be dictated by local law and organizational requirements.

The final core member of the HUET instructional team is the hoist operator. Instructors and divers are responsible for trainee safety in and on the water. The hoist operator is responsible for completing daily preuse checks, for ensuring the simulator is properly connected to the crane that is suspending it above the water, and for testing the simulator and associated lifting system before anyone steps foot inside it. The instructor will then guide trainees into the simulator and the hoist operator will raise them to the desired height for the exercise. On the instructor's command the hoist operator will lower the simulator to or into the water. If the UES is being submerged then the hoist operator needs to be watching the surface of the water. The hoist operator will be counting how many trainees and staff are surfacing and she/he is also waiting for the divers to give an all clear prior to recovering the simulator and then positioning it for the next exercise. The other thing he/she is watching and waiting for is a diver to call an emergency raise. If a raise is called, the hoist operator will immediately raise the simulator approximately one-half of a meter out of the water creating an air gap large enough that personnel on the inside of the simulator can stand up and breathe. To ensure that a trainee will not fall headfirst to the floor if still secured by his/her seat belt and the seat belt is released, the UES is not automatically raised completely out of the water. If, following an emergency raise, more of an air gap is required, either an instructor or a diver can advise the hoist operator accordingly.

Performance Evaluation

The evaluation of trainee performance is just as critical as the delivery process. The goal of the training provider is to ensure, to the best of his/her ability, that trainees are prepared to survive a helicopter ditching and to survive at sea until rescued. Just as there is no unified HUET system for program delivery, there is currently no definitive evaluation process in place for HUET performance. Moreover, very little research has been conducted that provides evidence that successful completion of a HUET program has a positive effect on surviving a ditching (Cunningham, 1978; Ryack, Luria, & Smith, 1986). We can, however, conclude that completion of a HUET program improves confidence in egress skills, increases comfort in the water, and increases an individual's belief that she/he will be more likely to survive a helicopter in-water accident (Taber & McGarr, 2013).

A common approach for determining success during HUET programs is to have trainees perform required egress skills unassisted. Although this is certainly better than the instructor continually assisting the trainee, it is not a true evaluation of performance. By having the trainees demonstrate required skills immediately after they have been given instructions for the egress procedure makes it difficult to determine if they have retained those skills beyond a short-term interval. It has been recommended that trainees should have a minimum of 24 h between initial acquisition of a skill and testing of recalled performance (Savion-Lemieux & Penhune, 2005; Taber & McGarr, 2013). The time is known as the consolidation period. One-day HUET programs make the 24-h consolidation period difficult to manage logistically but programs that run over multiple days, such as an OPITO BOSIET or CAPP BST, may be able to assess performance and evaluate skills later in their respective programs.

Characteristics of an Effective HUET Program

Effective HUET programs should be designed and organized in such a manner that trainees are able to recognize when they are in danger, prepare for a ditching using whatever time they have available, and emphasize surviving the impact phase and minimizing injuries. The program should also prepare trainees to survive at sea with or without a life raft. Finally, it should make the trainees aware of the various types and opportunities for rescue and the general actions and considerations to ensure the rescue effort is a success.

Whenever possible trainees should be encouraged to train using the same equipment they would normally fly in and the UES should be as accurate a representation as possible (Kozey, McCabe, & Jenkins, 2007; see also Chapter 7). Emphasis for the simulator should be placed on seat height and distance, seating or equipment configuration, restraint harnesses, exit size, and functionality. It should also contain meaningful and easily recognizable physical reference points such as auxiliary fuel tanks, grab handles, overhead consoles, stretchers, or aircraft-mounted weapons, for example. The exercises should mimic real-world ditching scenarios that are applicable to that particular group, such as an offshore helicopter losing power during takeoff from an oil rig or a search and rescue aircraft running out of fuel in bad weather. The more relevant the training is to the trainee, the more likely that he/she will retain the information and skills into the future. Training providers should strive to create challenging but achievable results that increase the trainees' confidence in their ability to survive. They must also ensure that they can repeat the process and that the training can be conducted safely. Performance evaluations should be well thought out and, whenever possible, implemented at a date later than when the training was delivered. The entire program should be internally reviewed in an effort to keep it current and engaging. Above all else, the trainee must see the value in acquiring knowledge and participating in the training. Better to be a survivor than a statistic.

REFERENCES

Brooks, C. J. (2007). The human factors of surviving a helicopter ditching. In *Research and Technology Organization, North Atlantic Treaty Organization, Survival at sea for mariners, aviators, and search and rescue personnel. RTO AGARDograph, AG-HFM-152.*

Brooks, C. J., MacDonald, C. V., Donati, L., & Taber, M. J. (2008). Civilian helicopter accidents into water- analysis of 46 cases, 1979-2006. *Aviation, Space, and Environmental Medicine, 79*(10), 935–940.

Brooks, C. J., & Potter, P. L. (1998). The abysmal performance of the inflatable liferaft in helicopter ditchings. In *RTO HFM symposium proceedings. RTO MP-19.*

Canadian Association of Petroleum Producers. (2013). *The Atlantic Canada offshore petroleum industry: Standard practice for the training and qualifications of personnel (Standard Practice) 2013-0007 (Replaces Publication 2010-0028).* Retrieved from http://capp.ca/publications-and-statistics/publications/223065.

Cheung, S., D'Eon, N., & Brooks, C. J. (2001). Breath holding ability of offshore workers inadequate to ensure escape from ditched helicopters. *Aviation, Space, and Environmental Medicine, 72*(10), 912–918.

Cheung, B., Hofer, K., Brooks, C. J., & Gibbs, P. (2000). Underwater disorientation as induced by two helicopter ditching devices. *Aviation, Space, and Environmental Medicine, 71*(9), 879–888.

Civil Aviation Authority. (2013). *CAP 1034 – Development of a technical standard for emergency breathing systems.* Norwich, UK: TSO (The Stationary Office) on behalf of the UK Civil Aviation Authority.

Civil Aviation Authority. (2014). *CAP 1145 – Safety review of offshore public transport helicopter operations in support of the exploitation of oil and gas.* West Sussex, UK: Safety and Airspace Regulation Group.

Coleshaw, S. R. K. (2006). *Stress levels associated with HUET: The implications of higher fidelity training using exits.* Report SC 155; prepared on behalf of OPITO, Aberdeen. Retrieved from http://www.opito.com/library/documentlibrary/huet_stress_report.pdf.

Coleshaw, S. R. K., & Howson, D. (1999). Escape from side-floating helicopters. In *Paper presented at the second international helicopter escape seminar/workshop.* Billingham, UK.

Cunningham, W. F. (1978). Helicopter underwater escape trainees (9D5). *AGARD Conference Proceedings.*

Glendon, A. I., Clarke, S. G., & McKenna, E. F. (2006). *Human safety and risk management* (2nd ed.). Boca Raton, Fl: Taylor and Francis Group.

Knass, K. (2013). *How I survived North sea helicopter crash that killed four – Hull man Paul sharp.* Retrieved from http://www.hulldailymail.co.uk/Holderness-Road-man-escaped-death-helicopter/story-20093886-detail/story.html#ixzz2qaRpl0Qo.

Kozey, J., McCabe, J., & Jenkins, J. (2007). The effects of different training methods on egress performance from the Modular Egress Training Simulator. In *Safe conference proceedings.*

Leach, J. (2004). Why people freeze in an emergency: temporal and cognitive constraints on survival responses. *Aviation, Space, and Environmental Medicine, 75*(6), 539–542.

Mills, A. M., & Muir, H. (1999). *Development of a standard for underwater survival.* Technical Paper for Shell Group.

Norwegian Oil & Gas. (2013). *Recommended guidelines for safety and emergency preparedness training.* Translated version, version 21.

Oil & Gas UK. (2014). *UK Continental shelf offshore workforce demographics report 2014.* The United Kingdom offshore oil and Gas Industry association trading as oil & Gas UK. Document available at http://www.oilandgasuk.co.uk/knowledgecentre/Offshore_workforce_demographics.cfm.

OPITO. (2013). *Basic offshore safety induction & emergency training, helicopter underwater escape training and further offshore emergency training.* Approved Standard. Revision 5 Amendment 4 date: 29/01/2013. Retrieved from http://www.opito.com/media/downloads/bosiet-huet-foet.pdf.

Robinson, S. J., Sünram-Lea, S. I., Leach, J., & Owen-Lynch, P. J. (2008). The effects of exposure to an acute naturalistic stressor on working memory, state anxiety and salivary cortisol concentrations. *Stress, 11*(2), 115–124.

Ryack, B. L., Luria, S. M., & Smith, P. F. (1986). Surviving helicopter crashes at sea: a review of studies of underwater egress from helicopters. *Aviation, Space, and Environmental Medicine, 57*(6), 603–609.

Savion-Lemieux, T., & Penhune, V. B. (2005). The effects of practice and delay on motor skill learning. *Experimental Brain Research, 161*, 423–431.

Summers, F. (1996). *Procedural skill decay and optimal retraining periods for helicopter underwater escape training*. Willetton, Western Australia: IFAP.

Taber, M. J. (2010). Offshore helicopter safety inquiry report. Commissioner. In R. W. Wells (Ed.). *Canada-Newfoundland and Labrador offshore helicopter safety inquiry* (Vol II). p. 211–290.

Taber, M. J. (2013). Crash attenuating seats: Effects on helicopter underwater escape performance. *Safety Science, 57*, 179–186.

Taber, M. J. (2014). Simulation fidelity and contextual interference in helicopter underwater egress training: an analysis of training and retention of egress skills. *Safety Science, 62*, 271–278.

Taber, M. J., & McCabe, J. (2006). Helicopter ditching: time of crash and survivability. *SAFE Journal, 34*(1), 5–10.

Taber, M. J., & McCabe, J. (2007). An examination of survival rates based on external flotation devices: a helicopter ditching review from 1971 to 2005. *SAFE Journal, 35*(1), 1–6.

Taber, M. J., & McCabe, J. (2009). The effect of emergency breathing systems during helicopter underwater escape training for land force troops. *Safety Science, 47*(8), 1129–1138.

Taber, M. J., & McGarr, G. W. (2013). Confidence in future helicopter underwater egress performance: an examination of training standards. *Safety Science, 60*, 169–175.

Taber, M. J., & Bohemier, A. (2014). Aeromedical aspects of maritime helicopter operations: CH124 Importance in human factors survival training – surviving the unthinkable. In *Sic Itur Ad Astra series on Canadian Air Power History* (Vol. 5).

Wells, R. (2010). *Canada-Newfoundland and Labrador offshore helicopter safety inquiry* (Vol. I). St. John's, NL: Canada-Newfoundland and Labrador Offshore Petroleum Board.

CHAPTER 3

Psychological Factors in Underwater Egress and Survival

John Leach
University of Portsmouth, Portsmouth, UK

3.1 INTRODUCTION

In many aircraft accident inquiries an intuitive distinction is made between "mechanical failure" and "human error." While "failure" is a relatively neutral term, "error" suggests fault, which in a human context implies blame. Mechanical failures can be described quite specifically (e.g., engine failure in hover, oil pump failure, tail rotor control failure, fuel starvation, etc.). This level of description and analysis should also be applied to the malfunction of the cognitive system (e.g., processing failure in working memory, malfunction of executive attention, failure to recall required information from semantic memory, failure to recall emergency response from procedural memory, etc.). This is not as difficult as it may first appear and an understanding of the "mechanical" failures of the cognitive system during a rapidly unfolding emergency such as a helicopter ditching should increase survival rates.

Our behavior is planned, initiated, monitored, and modulated through our cognitive system. Information is collected from the surrounding environment through transducers (ears, eyes, etc.) and then transmitted to our brain through nerve impulses where a model is constructed of the world around us. Our cognitive system continues to monitor incoming information, which it compares to this dynamic model and updates our behavior accordingly. It is important to note that we respond to the model we have created of the surrounding environment and not directly to the environment itself. We assume that the model in our head is a fair reflection of the ground reality. For the most part this cognitive model is sufficiently accurate and robust, but occasions can arise during which the external world changes so suddenly that our cognitive model lags behind the new reality. It takes time to create a fresh representative model of the new environment and it is our behavior during this period that often determines

Handbook of Offshore Helicopter Transport Safety
ISBN 978-1-78242-187-0
http://dx.doi.org/10.1016/B978-1-78242-187-0.00003-1

whether or not we survive. It is our continuing to respond to the old model as though it were still valid that produces "human error."

As humans we live our lives through routine. This is necessary because our brain is a limited-capacity processor and routines require no intensive higher-order cognitive processing, thus releasing scarce cognitive resources to plan our short, medium, and long-term goals while also monitoring our immediate environment and carrying out our current tasks. Relying on routine patterns of behavior to support our day-to-day survival has proven to be a robust and resilient system. Our cognitive system predicts the immediate future, incoming information is mapped against our predictions, and any deviation between what we expect to happen and what is actually happening is usually amended with little fuss: an item you require is unavailable so you substitute with something else; the road ahead is blocked so you work out another route to home; the person you have arranged to meet is ill so you reschedule the meeting. These examples are inconvenient but not life-threatening.

An emergency arises when there is a threat to life and our intrinsic world of routine and prediction switches suddenly to an extrinsic environment of uncertainty and danger. This switch usually happens faster than the cognitive system can update itself to compute the changes and establish a new plan of behavior. As pertinent examples consider an airplane or helicopter ditching in water, a ship sinking at sea, or a bus plunging into a lake or river. The original flight plan, ship's course, or road route has been plotted within various degrees of exactitude and the pilot, master, or driver is responding to this unfolding plot. When the emergency occurs the predicted plot becomes redundant and a new response has to be created. The problem arises when the threat unfolds faster than the cognitive system can create a new and appropriate response.

3.2 THE BRAIN AS AN INFORMATION-PROCESSING SYSTEM

Our cognitive system can be viewed as a limited-capacity signal processor that has to operate in real time. There are three main components to this system that concern survivor behavior: working memory, executive attention, and the supervisory system.

3.2.1 Working Memory

Working memory holds and manipulates operational information over short periods of time. A simple example is that of mentally multiplying numbers together (e.g., 7 × 19) without the aid of pen or paper. Working memory combines and processes information obtained from the environment with stored information recalled from long-term memory. There are two basic subdivisions of working memory that operate more or less independently: verbal working memory and visual working memory (Baddeley, 2011).

As with any other information-processing device, working memory is limited in both its storage capacity and its processing capacity and consequently can hold and process only so much information at one time and no more, with the speed of processing being limited by the rate of travel of a nerve impulse (Newell, 1994). For example, mentally multiplying 7 × 9 is quite easy and by a simple extension 7 × 90 is not too difficult, whereas multiplying 72 × 937 is problematic for most people.

3.2.2 Executive Attention

Executive attention is a combination of selective and sustained attention that enables modulation of working memory capacity and serves to maintain operational information in an active and quickly retrievable state (Engle, 2002). Importantly, executive attention also inhibits unwanted distracting information, resolves conflicts between incompatible sources of information, and focuses cognitive resources onto important tasks. Executive attention is the functional link between the supervisory system and working memory and can be thought of as the guidance system for enabling goal-directed behavior.

3.2.3 The Supervisory System

The supervisory system enables goals to be formulated and supports executive function, which has been succinctly described as "… the ability to plan and co-ordinate a wilful action in the face of alternatives, to monitor and update action as necessary and suppress distracting material by focusing attention on the task at hand" (Jones & Harrison, 2001). This is important because survival requires goal-directed behavior.

The advantage of the supervisory system is that it enables us to establish goal-directed survival behaviors before they are needed and to store these behaviors in parts of our memory system where they can be activated when necessary and, importantly, when the supervisory system is no longer functional (i.e., when it is part of an organism that has just ditched into a cold, wet, and threatening sea).

Essentially, the supervisory system plans and initiates behavioral goals; executive attention selects and focuses on the information needed to support those goals and prevents distracting information from intruding; and working memory processes the selected online information to formulate the behaviors needed to realize those goals.

As with all other information-processing devices, the cognitive system will be constrained by the limited resources available to it. Being a limited-capacity system, the brain can process information only up to a maximum rate, and being in a life-threatening emergency does not alter this fact. Simple cognitive operations, such as selecting between a set of prelearned responses, can be made in 1–2 s, while a simple two-choice decision can be achieved in 0.5 s; however, for the supervisory system to create a new response or initiate an unplanned behavior takes between 8 and 10 s (Newell, 1994), and these times are for optimal processing conditions. It is not possible to think ahead in an emergency, in which events unfold faster than the cognitive system can process them, although it is often possible to initiate prelearned behaviors.

Learning serves to reduce the constraints of a limited-capacity system by converting a series of initially complex operations into a simple, sequentially correct schema of behavior that is then stored in the relevant memory system and can be activated by the appropriate stimulus trigger. This process underlies all learned behavior, from playing a musical instrument to escaping from a submerged helicopter. Once the response has been learned, the brain no longer requires deliberation or higher-order cognition to compose the correct schema but has only to select between a set of prelearned responses. One drawback is that this schema can be recalled and implemented only in the manner in which it was learned. To restructure the activities would involve decomposing simple cognitive operations back into their complex components with consequent reversion to loading on working memory capacity.

3.3 NATURAL HISTORY OF AN EMERGENCY

From a psychological perspective an emergency follows a defined sequence: preimpact, impact, recoil, rescue, and posttrauma (Leach, 1994). This chapter is concerned with the first three of these phases.

3.3.1 The Preimpact Phase

The preimpact phase incorporates the knowledge, training, and relevant experience an individual possesses to support his or her survival response to the emergency: swimming ability, helicopter underwater escape training (HUET), lifeboat drills, sea survival practice, training in coastal survival skills, etc. This knowledge, training, and skill set underpins the correct behavioral responses during an emergency.

Underwater evacuation training has to be considered but as one part of a survival process. How we behave during the preimpact phase determines how successfully we respond during the impact phase of an underwater escape. Psychologically, these two phases are interlinked. First, the individual has to actually attend an underwater escape training course (see Chapter 2 for further details). This requires overcoming the urge to deny all suggestions that your aircraft may not actually reach its scheduled destination because it has instead flown into the sea. Denial can reduce anxiety but it also reduces the likelihood of survival. Many people will go out of their way to avoid HUET training: it can be unpleasant, strenuous, and anxiety-provoking; it is not relevant to the individual's main job in life, and it costs time and money, particularly for civilian offshore workers who may have to pay for such courses themselves.

We are capable of denial right up to the moment of impact: not attending to preflight safety briefings; not checking life jackets, safety equipment, and escape routes; not doing up the immersion suit correctly and directing our attention instead to a book or newspaper. Denial of a ditching prospect can be overcome if the person asks himself or herself the question: "What is my first response?" This is a simple but effective question that triggers the supervisory system to organize the relevant cognitive components to provide an answer and in so doing also establishes a psychological state of preparedness for an emergency. As the mind works by association, the answer to "What is my second response" will often occur without having to pose the question. Having now accepted that a survival incident may actually arise, the threshold for triggering the correct response is reduced and the brain will access and prime the relevant HUET or other appropriate training. Of course, if there is no relevant training to access, then the brain will continue to run behavioral routines in its default mode or, alternatively, will do nothing.

3.3.2 The Impact Phase

The impact phase is when the threat strikes. It is characterized by being sudden, violent, uncontrollable, and often accompanied by an overwhelming of the senses. One report describes a loss of aircraft control in a helicopter that ditched in water spinning on its axis, followed by the helicopter fuselage rolling inverted left and right, with breaking or bending rotor blades (Brooks, 1989).

The impact phase usually lasts for a few seconds but may extend for a few minutes and can comprise multiple assaults upon the person; the physical striking of the helicopter upon the water, the inrush of water that has been likened to being hit in the chest by a fire hose (Brooks, 1989), the inversion of the fuselage with consequent jerking of the body, disorientation, and further physical and psychological shocks. It is often shorter for helicopter passengers, as their awareness of something being wrong usually comes after that of the aircrew.

3.3.3 The Recoil Phase

The recoil phase begins once the impact dangers have subsided and survivors begin to show a gradual return of awareness and cognitive function although not necessarily a full understanding—to know your helicopter has ditched is not always to know why your helicopter has ditched. The recoil phase may take up to three days before normal cognitive function and behavior are recovered (Leach, 2012). This delay can expose the victim to secondary threats such as a risk of hypothermia following sudden and unexpected immersion in cold water (see Chapters 4 and 9). Consequently, merely escaping from a submerged helicopter does not guarantee survival.

3.4 COGNITIVE DYSFUNCTION IN EMERGENCIES

All three main components of the cognitive system (working memory, executive attention, and supervisory system) are vulnerable to dysfunction under threat or environmental duress. Egress from a submerged aircraft is further complicated by a short time limit on breath-holding that can be lessened further in cold water (Haywood, Hay, Matthews, Overwheel, & Radford, 1984). The underwater survival time can be extended by an additional source of oxygen such as a compressed oxygen cylinder or a survival suit air pocket, enabling exhaled air to be rebreathed. Again, such operations need to be prelearned.

A comparable threat situation with strictly limited time for survivability is that of parachuting when a main canopy malfunction occurs and the parachutist has a short time in which to pull the emergency handle and deploy the reserve parachute. This does not always happen; for example, in 2004, a parachutist with 118 previous jumps died when, despite a malfunction of his main parachute canopy, it was found that he had not attempted to pull his reserve handle and the mechanism was found to be intact (www.dropzone.com/safety/). In 2005, a parachutist with over 2300 recorded jumps died when she failed to deploy her reserve parachute; the cutaway handle was still in its place and the whole system was found to be working properly but with no sign of its being activated. These are two examples, but still there are others: a study of 241 sport parachuting fatalities between 1993 and 1999 found that 26 (11%) were due to a failure to pull the reserve parachute deployment handles (Griffith & Hart, 2002). As an aside, it would be interesting to compare this figure of 11% with the percentage failure in deploying escape hatches under water, given that survivability is severely time limited in both situations. If the figures are comparable then this would suggest a generic failure in cognitive function.

A similar observation was made in a HUET study by Brooks, Bohemier, and Snelling (1994), who report that two pilots when under water forgot to use the technique that they had been taught to locate the escape hatch lever, and the instructors had to place the pilots' hands onto the lever. Afterward, the pilots reported having no difficulty with either locating or operating the lever. Other personnel also reported having no difficulties when under water despite requiring intervention and assistance by instructors to the extent that they would not have survived had it been a real accident. This suggests that under water there is difficulty in both recalling essential information and encoding new information. The implication from these cases is that HUET training should be balanced between simulating ditching conditions realistically and maintaining low levels of stress or anxiety. The higher the anxiety level in the student, the less he or she will learn (see Chapter 6). Similarly, parachutists are often unable to recall what happened during their first jumps (Breivik, Roth, & Jorgensen, 1998) and poor learning of word lists under "in-air" conditions compared to learning on the ground has been found (Thompson, Williams, L'Esperance, & Cornelius, 2001), which suggests that poor encoding occurs under extreme emotionally arousing circumstances.

To investigate cognitive impairment under threat, the storage and processing capacities of novice and experienced parachutists were compared under three conditions: prior to exiting the aircraft, on landing, and during a control, nonjumping day (Leach & Griffith, 2008). Analysis showed that both experienced and novice parachutists had significant and comparable impairment in both storage and processing capacities prior to exiting the aircraft (at maximum danger perception), while on landing, the experienced parachutists showed recovery to normal storage levels, unlike the novices, who continued to show a delayed restriction in storage capacity; both groups had recovered normal processing capacity immediately after landing. The study also showed that the parachutists had difficulty in recalling information from long-term memory. A separate study examined working memory capacity in people undergoing HUET and again found impairment in working memory function when tested immediately following egress compared with before and after periods (Robinson, Sunram-Lea, Leach, & Owen-Lynch, 2008).

Executive attention has also been found to be vulnerable to environmental duress. Selective attention, sustained attention, attentional switching, and auditory–verbal working memory were studied over five days during a military aircrew survival exercise that simulated an "aircraft down" emergency. Compared to both a baseline condition and a control group, those undergoing environmental duress showed significant impairment in executive attention. No impairment was found in either attentional switching or auditory–verbal working memory. Impairment in executive attention makes flexible interactions with the survival environment difficult, and the person's behavior becomes dominated by environmental cues at the expense of willful, goal-directed survival behavior. Recovery to normal functioning was not immediate and in some cases it took two to three days for normal functional levels to return (Leach & Ansell, 2008). This has implications for the recoil period when, after having escaped from the submerged aircraft, the victim then has to cope with survival aboard a life raft or possibly on land after being washed ashore.

Supervisory system function was examined during another intensive survival exercise involving military aircrew (Porter & Leach, 2010). Supervisory system function is not a unitary process, but rather an integrated yet separable system comprising at least five executive subprocesses, which include the capacity to coordinate simultaneous activities; the ability to switch from one mental procedure to another; the suppression of prepotent responses; the capacity to encode (learn), access, retrieve, and manipulate information in long-term memory; and the ability to plan ahead (Baddeley & Della Sala, 1996; Carpenter, Just, & Reichle, 2000). Individual measures were taken of each of these processes over five days of the survival exercise and the results showed that, while undergoing the survival field phase, participants had significant impairment in inhibition, set switching, and planning ability. No significant impairments were found in dual-task performance or in the recall of total items from long-term memory over the five days, although the strategy for information recall changed.

3.5 SURVIVOR BEHAVIORS: DYSEXECUTIVE SURVIVOR SYNDROME

Some behaviors may be maladaptive but they are not haphazard. As different components of the cognitive system fail to cope with the sudden change from an environment of routine to one of threat, the incoming environmental information is processed in a way that produces behaviors that have an internal logic but are externally maladaptive (Leach, 2012). The specific types of behavior frequently witnessed during a survival incident comprise loss of initiative, stereotypy, perseveration of thought and action, hyperkinesia, hypokinesia, and, in extreme cases, akinesia or cognitive paralysis. Together these behaviors comprise a dysexecutive survivor syndrome (Leach, 2012).

3.5.1 Loss of Initiative

A common initial response to threat is an inability to think or plan clearly. One survivor of the ferry *MV Estonia*, which sank off Finland in 1994 with the loss of 852 lives, reported: "I didn't think. Shock is so disorienting it doesn't allow us to think clearly" (Leach, 2004). The explosion and collapse of the *Piper Alpha* oil platform in 1988 resulted in the deaths of 167 men. The official inquiry found that the death toll was considerably greater than it would have been if initiative had been taken with respect to escaping (Cullen, 1990). Furthermore, "Individual actions are not encouraged on oil platforms but in this case it was those who took matters into their own hands who survived. The rest did not" (Cook, 1989).

The pilot of a light aircraft that ditched and ended up inverted in the water following an engine failure describes his experience as follows: "My first instinct was to attempt pounding out a window, but water resistance prevented me from developing sufficient force. Panic began to set in, and I felt pain developing in my chest as I resisted the natural and almost overwhelming urge to inhale. After conceding that this might be the end of my life, the thought finally occurred to open the door. I did and found myself standing in five feet of water. The panic caused by suddenly discovering myself inverted and under water in an airplane had numbed my mind and delayed logical action almost until it would have been too late" (Schiff, 2001).

3.5.2 Stereotypy

Stereotypical behavior has the advantage of being behavior that has become automatic and routine. It also has the disadvantage of being behavior that is initiated without reflection. For example, aircraft evacuation testimonies report that passengers, instead of exiting the aircraft immediately, will stop to retrieve their carry-on baggage from the overhead bins despite instructions not to do so (TSBC, 2008). As an example, during a Boeing 737 engine fire the aircraft was stopped on the runway and an evacuation ordered. The crew stated, "Despite being advised to leave everything behind, many passengers insisted on retrieving their carry-on baggage. When confronted at the exits by cabin attendants, some passengers tried to return to their seats to stow their baggage in the overhead bins" (AAIB, 1990).

Why did they do this? Because this is what they always did before leaving the aircraft; the fact that this was an emergency with an intrinsic threat to life had not registered. Their cognitive model of the new environment had not been updated.

Stereotypical behavior can be an inherent response (instinctive) or it can be acquired through training (second nature). Focused and directed training can utilize stereotypy to provide an appropriate response to a sudden threat, for example, HUET instills behaviors that can be initiated reflexively (viz. stereotypically) to aid a person in surviving a ditched and submerged helicopter. It is clear that establishing adaptive responses through training such as HUET works. In a study of US helicopter ditchings (1963–1975) the survival rate among those with underwater helicopter escape training was 91.5%, but without training survival was 66% (Cunningham, 1978). An important part of this training is learning to inhibit more prepotent behaviors, such as not inflating the life jacket while inside the fuselage, or exiting the aircraft while the rotor blades are still turning above the victim, or attempting to disengage the restraining belt as though it were a more familiar car seat belt, which has been observed on various occasions during HUET training (see also the Chapters 6 and 7).

3.5.3 Perseveration

This refers to the continuous repetition of a routine behavior irrespective of its fitness to the changed environment. Perseverative behavior was observed in the action of a woman passenger caught in a fire at Manchester airport (1985) who, to aid evacuation, was told to open the emergency over-wing exit door positioned next to her seat. The opening of the door involved lifting a lever upward, but the woman was seen instead to be pulling repeatedly on the armrest of her seat (AAIB, 1988). Similarly, in a HUET study examining escape procedures under water Brooks et al. (1994) report several people continuing to push furiously on the hatch escape lever even after the window had jettisoned.

Perseverative behavior exists in two forms: continuous and recurrent. In continuous perseveration the person produces immediate repetitions of actions, whereas in recurrent perseveration the person breaks out of the action sequence but resumes it after an interval. Perseverative behavior does occur in survival situations, although observations to date suggest that it is of the continuous type; that is, once a victim has broken out of a perseverative action he does not appear to return to it. Continuous perseverative behavior could also be identified during a study of the ergonomics of helicopter escape hatch jettisoning (Brooks et al., 1994) in which it was observed that if subjects who were submerged in a HUET simulator could not find the hatch jettison lever they would search "... aimlessly around the upper half of the window/door frame," thus showing perseveration of an action that the person appeared unable to inhibit; "However, once the escapee's hand was placed on the handle by one of the underwater observers, he did not have great difficulty in operating the jettison lever ...," which shows that the perseveration was of the continuous type. Once the maladaptive behavior of aimlessly searching was inhibited by an external source, the person did not revert to aimless searching.

3.5.4 Action Disorders

Three frequently reported abnormal action behaviors observed in survivors are hypoactivity, hyperactivity, and cognitive paralysis. The first two are really observable only during the recoil phase; however, the underlying behavior should be considered, as it can be relevant to survivorship following the impact of initial submergence, for example, having to survive in a life raft after escaping from a ditched helicopter.

3.5.4.1 Hypoactivity

This term refers to a decrease in both cognitive and motor (hypokinesia) functions. Victims will be passive and possibly appear depressed, demotivated, apathetic, and mentally withdrawing from the situation in which they find themselves. A good illustration of this behavior is the report of victims of the sinking of a cruise liner who were observed during the incident by two psychiatrists on board the rescue ship. The survivors acted initially as if they had been sedated, and the psychiatrists observed that the "... survivors presented themselves for the most part as an amorphous mass of people tending to act passively and compliantly. They displayed psychomotor retardation, flattening of affect, somnolence, and, in some instances, amnesia for data of personal identification. They were nonchalant and easily suggestible" (Friedman & Linn, 1957). In a separate incident an officer in charge of a lifeboat launched after the capsize and loss of a cargo ship logged that after a while, "... a sort of coma appears to have settled on to many of the crew, and whatever was done was done more or less automatically and without thought" (Foster, 1934).

Clearly apathy is maladaptive in a survival situation that requires goal-directed behavior and, interestingly, it appears not to be related to depressive disorders (Levy & DuBois, 2006). In other words, the observation of apathetic behaviors seen in some survivors is not related to clinical depression. This view is supported in a study by the author that investigated depressed behavior among a group of personnel undergoing a harsh winter survival exercise in which apathy, withdrawal, and depressive type reactions were clearly observed in some individuals. No evidence of clinical depression was identified over a three-day period from such measures as speech rate, word frequency, and Beck's depression inventory despite clear observation of depressed reactions.

3.5.4.2 Hyperactivity

This is seen as an intense but undirected liveliness or restlessness (hyperkinesia). Victims are prone to distraction and often seen flitting from task to task; for example, in the rapid sinking of a tug in the Gulf of St. Lawrence in midwinter, two very experienced Master Mariners paced backward and forward from one wing of the bridge to the other and did nothing as the water rose above their knees. They drowned, yet a much younger and less experienced deck hand standing next to them quickly donned his survival suit and survived (Brooks, 2008). Following an attack on shipping in the North Atlantic during World War II, approximately 300 men were rescued from open boats after 52 h afloat. The naval surgeon described the survivors' behavior thusly: "Everyone was in excellent spirits, but this amounted in several cases to mild hysteria, as shown by a temporary excitement and garrulousness" (hyperactivity) (Critchley, 1943).

3.5.4.3 Cognitive Paralysis

Of the various action disorders, cognitive paralysis leading to "freezing" behavior or catatonia in the face of danger is the most serious, as it prevents any survival response during the impact phase of the incident (Leach, 2005). Common speech describes such behavior in terms such as "struck dumb," "petrified," and "frozen stiff." The victim often fails to respond to even simple commands; for example, the official inquiry into the *MV Estonia* sinking reported that many people were "... passive and stiff, despite reasonable possibilities for escaping" (JAIC, 1997). A survivor of the *Estonia* sinking commented on his fellow passengers, "People just sitting in complete shock and me not understanding why they're not doing something to help themselves. They just sat there and being swamped by the water when it came in." In another incident, during an attempted aircraft evacuation, the passenger sitting next to the emergency exit made no attempt to open the door. It is reported that, even when directed by the cabin attendant to open the exit, she "... did not respond" (TSBC, 1986). The following describes a separate aircraft evacuation incident: "I asked one guy to open a door and he wouldn't, he just stood there ... I told another fellow 'open that door' ... after the hatch was opened he just set it down, right in the doorway, he didn't bother getting out of the window...he just stood there with the door open and the wind blowing in and the snow blowing in and I said to myself, that's the last straw, if nobody is going to move, I am" (TSBC, 1986).

During an aircraft fire that resulted in 55 fatalities, passengers were reported as sitting immobile in their seats until overtaken by smoke and toxic fumes (AAIB, 1988). Interestingly, an empirical study prompted by this incident observed some volunteers in an airliner simulator as being "behaviourally inactive" (Muir, Marrison, & Evans, 1989). Such freezing behavior occurring in an aircraft filling with smoke can equally apply to one filling with water.

This "freezing" is shown as an inability to move physically or to switch mentally from one set (routine existence) to another set (threat environment) or to generate new rules and behaviors to cope with and adapt to the changed environment. However, such instinctive "freezing" behavior can be adaptive under certain circumstances as it corresponds to a physical immobility coupled with a state of hypervigilance or a guarded alertness that seeks to detect movement in the environment (Bracha, 2004). Consequently, when faced with danger the body prepares not for one of two responses, "fight or flight" (Cannon, 1929), but for three and in the order "freeze, flight, or fight" (Leach, 2004). The adaptive "freezing" behavior sees a victim immobilized before a roaming predator becomes maladaptive when it causes victims to remain immobilized in their seats in a submerging helicopter.

Threat diminishes the availability of cognitive resources and restricts attentional capacity that hinders people from interacting flexibly with their environment in a goal-directed manner. This makes it difficult either to maintain task-relevant information in an active state or to suppress unwanted environmentally triggered stimuli from entering working memory. When these goal states are not actively maintained by working memory then behavior becomes disorganized, perseverative, or otherwise maladaptive. This restriction in working memory means that both the amount of information that can be held is reduced and the rate of processing is slower. Facing threat stimuli means that a person cannot think as well or as fast as he or she would normally: at best thinking is sluggish, at worst it ceases altogether.

Impairment in executive attention is known to occur under duress (Leach & Ansell, 2008) and is characterized by an overreaction to environmental stimuli and a difficulty in focusing and maintaining attention when faced with a surge of new environmental information. Such impairment makes it difficult for the person to select a target of focus (e.g., a seat belt release buckle, aircraft door handle, etc.) and to maintain focus on the target without being distracted and having attention captured by other irrelevant incoming stimuli. This results in stereotypy and perseveration of thought and action; in other words, in an emergency people continue to do what they always do until they realize that it is now an emergency. Continually switching to and from the target results in hyperactivity.

Behavioral responses are triggered by environmental stimuli. If an appropriate response has been created and stored within the brain and is available for recall during an emergency, then the chances of survival are increased (consider someone in a ditched helicopter who has undergone HUET training). If a response is available that is a close enough match to the incoming environmental stimuli, then this response can be triggered and acted out even if it is not adaptive or supportive of survival (consider victims of a crashed airliner removing their bags from the overhead bins before exiting the aircraft or pestering the purser for money while the ship is sinking, as occurred on the *MV Estonia*).

As well as impairment in executive attention disruption of the supervisory system is also known to occur under duress (Porter & Leach, 2010). This appears paradoxical given the unique role the supervisory system plays in the adaptation of humans to surviving in a dynamic and frequently novel environment (Burgess, 1997; Shallice & Burgess, 1996). However, the supervisory system is resource intensive, requiring significant amounts of time and energy to process information. Consequently, the role of the supervisory system in increasing survival rates must be more subtle than operating simply as a real-time cognitive processor when under threat.

It has been proposed that the reason for the inoperability of the supervisory system under threat is due to temporal constraints (Leach, 2004). As mentioned above, it takes 8–10 s to create a temporary schema of behavior, so a dangerous event unfurls faster than that cannot be processed by the supervisory system. If the supervisory system becomes impaired, then it cannot construct a new and appropriate schema of behavior to respond to the danger. Consequently, if the supervisory system has already formulated an appropriate response, then this response will be activated by matching environmental and perceptual triggers (e.g., in the event of a helicopter ditching in open water, a person who has successfully undergone HUET will initiate this escape response in the correct manner while bypassing the supervisory system). If no such prelearned behavior exists, and the supervisory system is unable to create one in time, then either an inappropriate response would be elicited, resulting in stereotypical or otherwise maladaptive behavior, or no response is triggered, resulting in cognitive paralysis or freezing.

In other words, if the supervisory system cannot create an escape plan, because of dysfunctional impairment or lack of time, and no other appropriate behavioral response can be triggered, then the victim will show paralysis in the face of danger. Such behavior was witnessed during the sinking of the *MV Estonia* in 1994: "Many [victims] were seen just holding on without moving; yet others appeared paralyzed"; "Some … people were beyond reach and did not react when other passengers tried to guide them, not even when they used force or shouted at them"; "Some were [just] sitting in corners, incapable of doing anything" (JAIC, 1997).

It has been argued that the supervisory system is responsible for processing information under threat conditions (Burgess, 1997), yet witness observations (Leach, 2004) suggest that supervisory system processing is actually vulnerable under threat and can easily become disrupted. Clearly, the supervisory system does have a role in survival but not during the period of impact, for which it is functionally unsuitable. Rather, the role of the supervisory system is to assist human survival by providing an increased temporal span, enabling individuals to model emergencies mentally before facing them for real. Indeed, it has been proposed that the reason our memory system has evolved in the way it has is to model the future rather than merely storing the past (Klein, Robertson, & Delton, 2010). In doing so, the supervisory system provides a buffer between us, our environment, and future eventualities. Once the need to model a potential emergency as a goal-directed activity has been accepted, the supervisory system enhances survivorship through practice, training, and experience in advance of any disaster.

However, it is clear from accident inquiries that this ability is not instinctive. Indeed, the response of many people is the opposite to that which intuitively would be expected; people will go out of their way to avoid considering any increased risk to their lives—the "don't even think about it" reaction. The superstition that contemplating a disaster will bring about its occurrence is strongly ingrained.

3.6 THE DITCHING: IMPACT PHASE

In normal flying, the pilot has planned the route and is working to a flight plan. He is using his supervisory system to enact goal-directed behavior to arrive at the destination. His executive attentional system is monitoring the external and internal environments to compare incoming information against the predicted plan and to bring into focus any alerts (e.g., the presence of other aircraft traffic in the vicinity, unexpected changes in instrument readings, etc.). Working memory capacity is deployed on processing operational information to ensure his behavior is aligned with the immediate operational goals established by the supervisory system. The cognitive system is functioning as a normal servo-feedback mechanism. Then the aircraft ditches.

This is the psychological impact phase, which usually lasts only a few seconds, although it may extend to a minute or so. A study of 46 Canadian helicopter accidents occurring over water (1979–2006) found that 83% had a warning time of less than 15 s; 11% had a warning time between 16 s and 1 min, and 2% had a warning time of over 1 min. The warning times of the remaining 4% could not be estimated (Brooks, MacDonald, Donati, & Taber, 2008). In 2013, a Super Puma L2 helicopter carrying 16 oil workers and two crew ditched off the Shetland Isles near Scotland, killing four passengers. The RNLI (coastguard) rescue coordinator reported to the BBC, "There appears to have been a catastrophic loss of power, which meant the helicopter suddenly dropped into the sea without any opportunity to make a controlled landing. " One survivor said that the helicopter seemed to have a big power failure and just dropped into the sea, and that there was no time to brace.

As well as little warning time in a helicopter ditching, other factors have been identified that can compromise survival (Brooks, Muir, & Gibbs, 2001): intense disorientation through sinking, fuselage inversion and darkness, inrushing water and difficulty in breath-holding exacerbated by the physical force of water pressure and a cold-inducing gasp reflex, and the ease or difficulty of locating and operating the door or window jettison mechanism.

The victim has to locate the hatch jettisoning mechanism in poor visibility and often without any visual reference points. This requires access to a mental map and assumes that the person has already stored the location information (i.e., during a preflight briefing). Nothing can be achieved until the jettisoning device is within the survivor's manual work space, viz., that he or she can reach the handle. This means that the survivor has to move toward the jettison handle or the handle has to move toward the victim. This is a matter of engineering design; for further discussion of this topic see Brooks (1997) and Brooks and Bohemier (1997).

The difficulties in disorientation were observed in a HUET study, which while investigating the ergonomics of jettisoning helicopter escape hatches under water found that even highly experienced HUET instructors had difficulty in locating the escape devices. An apparently simple task when upright on land became extraordinarily difficult when inverted under water and hampered with excessive buoyancy and without gravitational references (Brooks et al., 1994).

The problem of disorientation when trapped under water in a helicopter is further confounded by many different designs of escape mechanism; they vary in design, size, number, function (pull, push, rotate, or tear away), location, and even if they are situated on the escape hatch itself or on the main fuselage but separate from the hatch that is being jettisoned. The view of Brooks et al. (1994) is that helicopter escape hatch jettison mechanisms have been poorly designed for underwater escape. One might go further and question whether jettison mechanisms have been designed at all for use under water, the focus rather being on the design of a handle that opens a hatch rather than recognizing the jettisoning device as an integral (hand–handle) locking component in a survival process. At the basic level the problem concerns designing a system that requires a free-floating body to identify, locate, move toward, engage, activate, and manipulate a mechanical device. It is a problem that in essence is not dissimilar to that encountered by a spacecraft docking mechanism, and perhaps design issues could be improved by approaching the jettisoning mechanism as a locking-on or docking design problem between two dynamic bodies.

Valuable time must also be expended in reorienting a cognitive map even if the physical relation between the person and the jettison mechanism remains the same. Studies have shown that the time taken to reorient mentally a three-dimensional object that has been rotated or inverted from its original position is linearly proportional to the extent of the rotation. That is, the more the object has been moved from its original position the longer it takes the person mentally to rotate the object back to its original position (Shepard & Metzler, 1971).

It has been pointed out that more people perish in helicopter ditchings because they could not evacuate the sinking aircraft than through injury or physical ditching (see Chapter 8 for a discussion on emergency breathing systems). This might be stated as more people perishing in helicopter ditchings through cognitive failure than through injury or physical impact. A failure to evacuate should be considered from the cognitive perspective and, for anyone flying over water, the "ditched environment" should not be an unknown one. For the cognitive system to function effectively the environment has to become familiar.

3.7 POSTDITCHING: RECOIL PHASE

Successful evacuation of a ditched helicopter is only one component of survivorship. It has been suggested that approximately 50% of those who successfully evacuated from a ditched helicopter perished before rescue arrived (Brooks, 1997). These deaths occurred during the psychological recoil phase and this remains a key period of cognitive dysfunction.

During the recoil phase, working memory returns to normal capacity for both storage and processing, enabling efficient processing of operational information to be resumed. Executive attention may be slower in recovering and may, in some survivors, take up to three days for full functionality to return. This slow recovery adversely affects the supervisory system, which can have difficulty in establishing appropriate, adaptive, and goal-directed behaviors to aid survival. In the immediate aftermath of a ditching, a malfunctioning cognitive system can expose the survivor to secondary threats such as hypothermia or heat stress and can result in a failure to carry out survival drills with respect to life jackets, life rafts, the location and preparation of flares, the taking of sea-sickness tablets, and so on.

Survival requires goal-directed behavior, and an inability to establish relevant goals means that behaviors are not produced to support survival. This lack of adaptive, directed behavior can be seen as a lack of motivation or apathy. It is not uncommon to observe people simply huddled up and unresponsive in a life raft even during sea survival training exercises. As an example of general cognitive malfunction, consider the case of the sinking of a 176-foot fishing vessel 600 miles off Hawaii (Leach, 1994). The eight-man crew aboard the vessel split into two life rafts, which were tethered together by a 30-m rope and remained so during the voyage. The men on board in the two life rafts went with their separate psychological ways. One raft came immediately under the firm leadership of the third mate. In the other raft, which contained the ship's captain, leadership appeared to be stillborn. The third mate insisted that each person stand a watch both day and night and decided when flares would be fired. Food and water were inventoried, rationed, and issued to each man publicly. The raft was mopped regularly to keep it as dry as possible to prevent saltwater sores developing. This was in contrast to their sister raft, in which the crew failed to carry out any similar procedures and sank quickly into despair. Instead of discipline there was disintegration. Two weeks after the sinking of the ship they were rescued by a naval vessel. It was noted that the crew of the third mate's raft were able to physically climb the boarding ladder, while those in the other raft had to be brought on board in litters. The captain died the day before rescue. It is important to realize that both rafts started off equal in types of craft, food, and water (Leach, 1994).

3.8 CONCLUSION

It has been a historical convention to consider the human as a separate add-on to any mechanical or information-processing system. This conventional thinking has led to systems being designed to operate within a normal environment and responding to normative behavior. The difficulty lies in the fact that once the system is thrust into an abnormal environment, normative behavior becomes maladaptive and may indeed be counter-productive. In worst cases life is lost.

Everyday behavior serves to support our short-, medium-, and long-term goals; people board an aircraft for a reason. This goal-directed behavior is achieved by the interaction of three closely interlinked cognitive systems: the supervisory system, which enables executive functioning such as planning, goal formation, initiating willful action, and the monitoring, updating, and, where necessary, the inhibition of current, ongoing behaviors; executive attention selects the information needed to support those goals and prevents unwanted information from intruding; and the working memory processes the selected operational information to produce the appropriate behavior to realize those goals.

A sudden change in the environment or situation impedes the processing and modulation of cognitive information. If this new situation is also threatening, then such cognitive impediment can be exacerbated. Under a fast-moving threat, the supervisory system is too slow to process information in real time. Executive attention becomes impaired and may take hours or even up to three days in some instances for full operational recovery. This leaves the cognitive system open to responding to environmental stimuli rather than being able to initiate planned adaptive (life-saving) behaviors. Working memory capacity for both information storage and processing becomes depleted, with the result that the amount of operational information that can be held at one moment is reduced and the rate at which information can be processed is diminished.

The behaviors that commonly result from this multilevel impairment are loss of initiative, stereotypy, perseveration of thought and action, hyperactivity, hypoactivity, and akinesia, or cognitive paralysis. This constellation of behaviors occurring within a survival situation has been identified as dysexecutive survivor syndrome (Leach, 2012).

Three approaches should be made to aid survival:

First, the threat environment has to be normalized as far as possible through exposure and familiarity, and the correct (adaptive) behaviors have to become normative through training and experience. The end result is that when a helicopter ditches, the stimuli from this new, sudden, and threatening environment trigger the appropriate and adaptive responses to aid survival while bypassing the supervisory system.

Second, engineers should design helicopter emergency egress facilities to be operable in a threat environment that is submerged and inverted. It is recommended at least that the jettison system be designed to support a human locking-on system and, given the limited behavioral responses available under threat, there should be a move as far as possible toward standardization of jettison mechanisms across aircraft types.

Third, to overcome initial denial during the preimpact phase and to prime the appropriate emergency behavior when in a risk environment, each person should ask himself or herself: What is my first response? The rest will follow.

To finish with an analogy: in a study of helicopter ditchings Brooks et al. (2008) comment on the survival equipment and life rafts that were carried on board an aircraft. They note that, "… the life-rafts were incorrectly stored, and inadequately maintained and serviced." I would add that along with survival equipment is survival knowledge, and to function properly this knowledge also has to be correctly stored, adequately maintained, and regularly serviced.

REFERENCES

AAIB: Air Accident Investigations Board. (1988). *Report on the accident to Boeing 737-236, G-BGJL, at Manchester Airport on 22nd August 1985*. London: AAIB.

AAIB: Air Accident Investigations Board. (1990). *Report EW/C1174*. London: AAIB.

Baddeley, A. (2011). Working memory: theories, models, and controversies. *Annual Review of Psychology, 63*, 1–29.

Baddeley, A., & Della Sala, S. (1996). Working memory and executive control. *Philosophical Transactions of the Royal Society of London B, 351*, 1397–1403.

Bracha, H. S. (2004). Freeze, flight, fight, fright, faint: adaptationist perspectives on the acute stress response spectrum. *CNS Spectrums, 9*, 679–685.

Breivik, G., Roth, W. T., & Jorgensen, P. E. (1998). Personality, psychological states and heart rate in novice and expert parachutists. *Personality and Individual Differences, 25*, 365–380.

Brooks, C. J. (1989). *Human factors relating to escape and survival from helicopters ditching in water*. Neuilly, France: NATO AGARDograph No. 305 (E).

Brooks, C. J. (1997). Surviving a helicopter ditching: an engineering challenge. In *AGARD symposium on future aerospace technology in the service of the alliance, CP-600* (Vol. 1). France: Ecole Polytechnique.

Brooks, C. J. (2008). *Survival at sea for mariners, aviators and search and rescue personnel*. Brussels: NATO Research and Technology Organisation.

Brooks, C. J., & Bohemier, A. P. (1997). Helicopter door and window jettison mechanisms for underwater escape: ergonomic confusion! *Aviation, Space & Environmental Medicine, 68*, 844–857.

Brooks, C. J., Boheimer, A. P., & Snelling, G. R. (1994). The ergonomics of jettisoning escape hatches in a ditched helicopter. *Aviation, Space & Environmental Medicine, 65*, 387–395.

Brooks, C. J., MacDonald, C. V., Donati, L., & Taber, M. J. (2008). Civilian helicopter accidents into water: analysis of 46 cases, 1979–2006. *Aviation, Space & Environmental Medicine, 79*, 35–40.

Brooks, C. J., Muir, H. C., & Gibbs, P. G. N. (2001). The basis for the development of a fuselage evacuation time for a ditched helicopter. *Aviation, Space & Environmental Medicine, 72*, 553–561.

Burgess, P. W. (1997). Theory and methodology in executive function research. In P. Rabbitt (Ed.), *Methodology of frontal and executive function*. Hove: Psychology Press.

Cannon, W. B. (1929). *Bodily changes in pain, hunger, fear and rage*. Appleton: Oxford.

Carpenter, P. A., Just, M. A., & Reichle, E. D. (2000). Working memory and executive function: evidence from neuroimaging. *Current Opinion in Neurobiology, 10*, 195–199.

Cook, J. (1989). *An accident waiting to happen*. London: Unwin Hyman Ltd.

Critchley, M. (1943). *Shipwreck survivors: A medical study*. London: J&A Churchill.

Cullen, L. (1990). *The public inquiry into the Piper Alpha disaster*. London: HMSO, Department of Energy.

Cunningham, W. (December 1978). Helicopter underwater escape trainer (9D5). In *NATO AGARD Conference Proceedings No 255*.

Engle, R. W. (2002). Working memory capacity as executive attention. *Current Directions in Psychological Science, 11*, 19–23.

Foster, C. P. T. (1934). *1700 miles in open boats: The story of the loss of the SS Trevessa in the Indian Ocean, and the voyage of her boats to safety*. Oxford: Houghton Mifflin.

Friedman, P., & Linn, L. (1957). Some psychiatric notes on the Andrea Doria disaster. *American Journal of Psychiatry, 114*, 426–432.

Griffith, J. D., & Hart, C. L. (2002). A summary of US skydiving fatalities: 1993–1999. *Perceptual and Motor Skills, 94*, 1089–1090.

Hayward, J. S., Hay, C., Matthews, B. R., Overwheel, C. H., & Radford, D. D. (1984). Temperature effect on the human dive response in relation to cold water near-drowning. *Journal of Applied Physiology: Respiratory, Environmental and Exercise Physiology, 56*, 202–206.

JAIC. (1997). *Final report on the capsizing on 28th September 1994 in the Baltic Sea of the ro-ro passenger vessel MV Estonia*. Finland and Sweden: Joint Accident Investigation Commission of Estonia.

Jones, K., & Harrison, Y. (2001). Frontal lobe function, sleep loss and fragmented sleep. *Sleep Medicine Review, 5*, 463–475.

Klein, S. B., Robertson, T. E., & Delton, A. E. (2010). Facing the future: memory as an evolved system for planning future acts. *Memory & Cognition, 38*, 13–22.

Leach, J. (1994). *Survival psychology*. Basingstoke: Palgrave Macmillan.

Leach, J. (2004). Why people 'freeze' in an emergency: temporal and cognitive constraints on survival responses. *Aviation, Space & Environmental Medicine, 75*, 539–542.

Leach, J. (2005). Cognitive paralysis in an emergency: the role of the supervisory attentional system. *Aviation, Space & Environmental Medicine, 76*, 134–136.

Leach, J. (2012). Maladaptive behaviour in survivors: dysexecutive survivor syndrome. *Aviation, Space & Environmental Medicine, 83*, 1152–1161.

Leach, J., & Ansell, L. (2008). Impairments in attentional processing in a field survival environment. *Applied Cognitive Psychology, 22*, 643–652.

Leach, J., & Griffith, R. (2008). Restrictions in working memory capacity during parachuting: a possible cause of 'no-pull' fatalities. *Applied Cognitive Psychology, 22*, 147–157.

Levy, R., & DuBois, B. (2006). Apathy and the functional anatomy of the prefrontal cortex-basal ganglia circuits. *Cerebral Cortex, 16*, 916–928.

Muir, H., Marrison, C., & Evans, A. (1989). *Aircraft evacuations: The effect of passenger motivation and cabin configuration adjacent to the exit*. UK Aviation Authority.

Newell, A. (1994). *Unified theories of cognition*. Cambridge, MA: Harvard University Press.

Porter, H., & Leach, J. (2010). Executive dysfunction in a survival environment. *Applied Cognitive Psychology, 24*, 41–66.

Robinson, S. J., Sunram-Lea, S. I., Leach, J., & Owen-Lynch, P. J. (2008). The effects of exposure to an acute naturalistic stressor on working memory, anxiety and salivary cortisol concentrations. *Stress, 11*, 115–124.

Schiff, B. (December 2001). *Ditching egress training. Equipped to survive*. AOPA Pilot Magazine.

Shallice, T., & Burgess, P. (1996). The domain of supervisory process and temporal organisation of behaviour. *Philosophical Transactions of the Royal Society of London B, 351*, 1405–1412.

Shepard, R. N., & Metzler, J. (1971). Mental rotation of three-dimensional objects. *Science, 171*, 701–703.

Thompson, L. A., Williams, K. L., L'Esperance, P. R., & Cornelius, J. (2001). Context dependent memory under stressful conditions: the case of skydiving. *Human Factors, 43*, 611–619.

TSBC. (July 14, 1986). *A86P4053: Report of B737-275 air incident, Kelowna, B.C.* Quebec: Transportation Safety Board of Canada.

TSBC. (2008). *SA9501: A safety study of evacuations of large passenger-carrying aircraft*. Quebec: Transportation Safety Board of Canada.

CHAPTER 4

The Physiological Responses to Cold-Water Immersion and Submersion: From Research to Protection

Michael J. Tipton
University of Portsmouth, Portsmouth, UK

4.1 INTRODUCTION

This chapter tells the story of the evolving understanding of the hazards to be faced by those confronted with an immersion in cold water, specifically an underwater escape from a ditched, inverted helicopter. It is a fascinating story that starts in the mists of time and eventually links scientific knowledge to hazard, and from there to technical solution. As such, it is a celebration of applied science, the discipline that embodies the application of fundamental concepts to the solution of practical problems. The story also provides an insight into how long it takes to change peoples' understanding, the conservative forces that strive to maintain the status quo, and the requirement for catastrophe to promote change.

4.2 BACKGROUND

Sometimes advances come about because of a confluence of information that already existed. In such cases the "original thought" is founded on the recognition of an association between previously unrelated factors. It is not unreasonable to argue that this was the case with the association of the initial hazardous responses to immersion in cold water, deaths on helicopter ditching, and the provision of emergency underwater breathing aids (EUBA). While each of these factors existed in isolation, it was as recently as the 1980s that they were linked.

Perhaps the first recorded reference to an independent role for cold in immersion deaths came with the description in 450 BC by Herodotus of the ill-fated seaborne expedition of the Persian general Mardonius. He states: "those who could not swim perished from that cause, others from the cold."

Handbook of Offshore Helicopter Transport Safety
ISBN 978-1-78242-187-0
http://dx.doi.org/10.1016/B978-1-78242-187-0.00004-3

Interestingly, the first use of underwater breathing bags can probably be seen as early as the ninth century BC in Assyrian bas-reliefs (Figure 4.1).

Sir James Lind (1765–1823) was a British Naval physician who noted that cold was associated with body cooling, muscular fatigue, impairment of consciousness, and subsequent drowning. He recognized the importance of rewarming to reverse the adverse effects of body cooling. Another physician, Dr James Currie (1756–1805), was a Scotsman with a medical practice in Liverpool. He is best remembered for his anthology and biography of Robert Burns and his medical reports on the use of water in the treatment of fever (Currie, 1805), which contain the first systematic record in English of experiments on the effects of cold water immersion on humans and clinical observations using a thermometer. His interest in cold water is thought to have arisen in December 1790 when he stood in a crowd, helpless and frustrated, watching as the crew of an American sailing ship, stranded on a sandbank in Liverpool harbor, struggling to survive in the 5 °C water. With time, unable to hold onto their ship, the crew fell into the water and drowned. Currie's findings included identification that cooling occurs faster in water than in air, that there is a relationship between water temperature and survival, and that cooling continues post-immersion. He also advocated the hot bath as a rewarming method.

Figure 4.1 Neo-Assyrian bas-relief circa 870 BC probably showing an incident described in Ashurnasirpal II's written accounts of swimmers in the river Euphrates near the enemy capital of Suru in the land of Suhi. Some have claimed this is the first evidence of re-breather-assisted diving using inflated animal skins. Others claim that it would be impossible to submerge with this amount of buoyancy and the swimmers are using the bladders for support (life-jacket rather than diving equipment); either way the connection to the mouth makes this amongst the first recorded evidence of the use of a re-breathing bag.

In 32,000 trans-Atlantic voyages up to 1912 there had only been 25 cases (0.0008%) in which either the ship or lives had been lost, with the number of fatalities totaling 148. This, in part, explained the lack of adequate safety provision; the risk was not regarded as that great. This perspective changed on April 15, 1912, when the *Titanic* sank, claiming 1503 lives. While many good things emerged from this tragedy, including the International Convention for the Safety of Life at Sea, the subsequent preoccupation with hypothermia was not among them.

Despite statements at the formal investigation, such as that of 17-year-old first class passenger Jack Thayer, that "The cold was terrific. The shock of the water took the breath out of my lungs..." or that 200 yards was about the maximum distance any survivor claimed to have swum and climbed into a boat, the clues suggesting hazardous short-term responses evoked by immersion were not recognized. Instead, the facts that most of those who died on the *Titanic* did not "go down with the ship," but drifted away in their life jackets, and that 300 dead bodies, found floating in their life jackets, were pulled from the sea the next morning by the crew of the Mackay–Bennett, led to the belief that hypothermia was the primary hazard to be faced on immersion in cold water. That this could happen was primarily due to the flat calm conditions in which the *Titanic* sank; this enabled the life jackets to keep the airway clear of the water and prevent the loss of control of breathing on immersion, mentioned by Jack Thayer, from resulting in drowning. The preoccupation with hypothermia was perpetuated during World War II, in which two-thirds of the 45,000 Royal Navy personnel who died did so in the survival phase, often in remote places. The provision of life jackets and floats reduced the chances of drowning but not of hypothermia.

The preoccupation with hypothermia persists and still has an impact on a range of areas, including the estimation of survival time and consequent search and rescue policies, assumptions about the cause of death on immersion, and the provision of protective equipment. So it was that the original personal protective equipment provided for those flying offshore in helicopters was an immersion suit and life jacket. However, the evidence was growing that other hazardous responses needed to be considered. This evidence was statistical, anecdotal, and scientific. The UK Home Office Report of the Working Party on water safety (1977) found that approximately 55% of the annual open-water immersion deaths in the United Kingdom occurred within 3 m of a safe refuge (42% within 2 m) and two-thirds of those who died were regarded as good swimmers. The statistics remain about the same today (Tipton et al., 2014). Anecdotal accounts of people succumbing quickly on immersion in cold water with cardiac problems or drowning were common enough for the term "hydrocution" to evolve. As early as 1884, Falk mentioned in a scientific report the respiratory responses to cooling of the skin of the hand. The report of the infamous experiments conducted at the Dachau concentration camp during World War II (Alexander, 1945) included reference to the initial hyperventilation seen on immersion in cold water.

Later, the scientific literature (see below) on immersion examined and described the initial responses to sudden cold-water immersion. It was this literature plus the anecdotal accounts from fatal accident enquiries that, in 1981, led Golden and Hervey to propose four stages of immersion associated with particular risk:

Stage 1. Initial responses—first 3–5 min.

Stage 2. Short-term immersion—5–30 min: neuromuscular dysfunction leading to physical incapacitation caused by cooling of superficial nerves and muscle.

Stage 3. Long-term immersion—30 min plus: hypothermia will not occur earlier than this in adults even in the coldest water temperatures.

Stage 4. Post-immersion—circum-rescue collapse: collapse in arterial blood pressure and cardiac failure occurring just before, during, or just after rescue (Golden, Hervey, & Tipton, 1994).

This remains the most valid categorization of the hazards to be faced by those immersed in cold water and provides the definitive framework for the understanding and interpretation of accidents involving cold water immersion (e.g., Table 4.1). Of most interest to this chapter and this book is Stage 1, the initial responses to immersion in cold water.

Table 4.1 Value of the four stages of immersion associated with particular risks categorization in the analysis of anecdotal accounts from fatal accidents

"...they died within seconds of entering the water."	Stage 1
"...he unfastened his seat belt but failed to exit the helicopter."	Stage 1
"...he visibly weakened and was unable to board the life raft."	Stage 2
"...one life raft occupant lost teeth trying to open the bag containing the bailer."	Stage 2
"...he lost his grip on the life raft and drifted away."	Stage 2
"...some of the occupants of the life raft became withdrawn, one became agitated and aggressive, then unconscious."	Stage 3
"...his condition deteriorated significantly during winching."	Stage 4

Statements from the *Herald of Free Enterprise* (1987), *Cormorant Alpha* (1992), and *Estonia* (1994) disasters.

In 1989 we reviewed the initial responses to cold water immersion in humans (Tipton, 1989) and used the term "cold shock," with "shock" relating to the stimulating, emotive aspect of the response rather than any reference to the medical definition of shock. In that review it was also concluded that "the cold shock response can result in the death or serious incapacitation of an individual long before general hypothermia develops."

The cold shock response comprises a range of cardiorespiratory responses initiated by a sudden reduction in skin temperature (Cooper, Martin, & Riben, 1976; Goode et al., 1975; Keatinge & Evans, 1961; Keatinge, McIlroy, & Goldfien, 1964; Keatinge & Nadel, 1965). Within limits, the magnitude of the cold shock response is related to the rate of change of skin temperature (temporal summation) and the surface area of the body exposed (spatial summation) (Tipton, Stubbs, & Elliott, 1990). In naked individuals the response peaks in water were between 10 and 15 °C (Tipton, Stubbs, & Elliott, 1991). The cardiovascular component includes an increase in heart rate, cardiac output, and blood pressure. The hazard associated with these responses has probably been underestimated, as most of the experimental studies on cold shock have been undertaken with young, fit, and healthy volunteers. Also, although the initial incapacitation may be caused by a cardiac problem, agonal gasps close to death may result in the aspiration of water and apparent drowning. Finally, some of the cardiac problems are electrical disturbances and not therefore detectable at postmortem, so other causes of death are sought.

During head-out immersion with young, fit, and healthy individuals cardiac arrhythmias are observed in about 1% of immersions. However, this percentage rises to approximately 82% (Datta & Tipton, 2006) if the face is immersed and a breath-hold is undertaken, that is, the situation that occurs during a helicopter underwater escape. Tipton, Gibbs, Brooks, Roiz de Sa, and Reilly (2010) measured the electrocardiogram of 26 young, fit, and healthy males undertaking five helicopter underwater escape training runs in water at 29.5 °C. The runs were separated by a minimum of 10 min (not dissimilar to standard helicopter underwater escape training), and each one required a breath-hold of approximately 10 s. Across all runs, the authors identified 32 cardiac dysrhythmias and arrhythmias (25%) in 22 participants; all but six of these occurred just after submersion (break of breath holding). Aerobic fitness appeared inversely associated with the occurrence of arrhythmias, with no arrhythmias occurring in individuals with a predicted aerobic capacity of greater than 3.88 L min^{-1} (44.7 mL kg^{-1} min^{-1}). Tipton et al. (2010) confirmed earlier findings (Tipton, Kelleher, & Golden, 1994) when they concluded that helicopter underwater escape training produces cardiac dysrhythmias and arrhythmias that are mostly supraventricular, asymptomatic, and probably of little clinical significance in young, fit, and healthy individuals. It is not known if this is the case with an older, less fit cohort of people or in those undertaking longer breath-holds in colder water.

The recent theory of "autonomic conflict" (Shattock & Tipton, 2012; Tipton et al., 2010) suggests that the high incidence of arrhythmias seen during submersions, particularly around the time of the release of a breath-hold, is caused by concurrent stimulation of both divisions of the autonomic nervous system. Normally the sympathetic and parasympathetic divisions of this system act reciprocally, but during submersion sudden and profound cooling of the skin initiates the cold shock response, including a sympathetically driven tachycardia (increased heart rate); while cooling of the oronasal region of the face evokes a "diving response," which includes a vagally induced bradycardia (decreased heart rate) (de Burgh Daly & Angell-James, 1979). It is this conflicting input to the heart from the two divisions of the autonomic nervous system that is thought to produce dysrhythmias and arrhythmias; these are normally asymptomatic and harmless but, in the presence of cofactors such as heart disease, can result in fatal arrhythmias (Shattock & Tipton, 2012).

The respiratory component of the cold shock response includes a "gasp" response and uncontrollable hyperventilation, which prevents breath holding (Tipton, 1989). On submersion in 10 °C water average maximum breath-hold times are about 5 s when normally clothed, rising to about 20 s in those wearing uninsulated immersion "dry" suits (Tipton, Balmi, Bramham, Maddern, & Elliott, 1995). Goode et al. (1975) measured a mean inspiratory gasp of 2 L on initial immersion in 28 °C water, rising to 3 L in water at 11 °C (Tipton, 1989). To put this into context, the lethal dose for drowning in sea water for a 70-kg individual is approximately 1.5 L (Modell, 1971). The various components of the cold shock response are presented in Figure 4.2.

Given the above, it should come as no surprise that the initial respiratory responses to immersion are now regarded as the most hazardous responses for those immersed in cold water, and especially if that immersion requires a breath-hold to exit a ditched inverted vehicle such as a helicopter. However, by the mid-1980s the evidence accruing in the scientific literature had not permeated through to the operational sector, and the routine provision of EUBA for civilian helicopter passengers was still some way off (see Chapter 8). This was despite a growing body of anecdotal evidence. In the

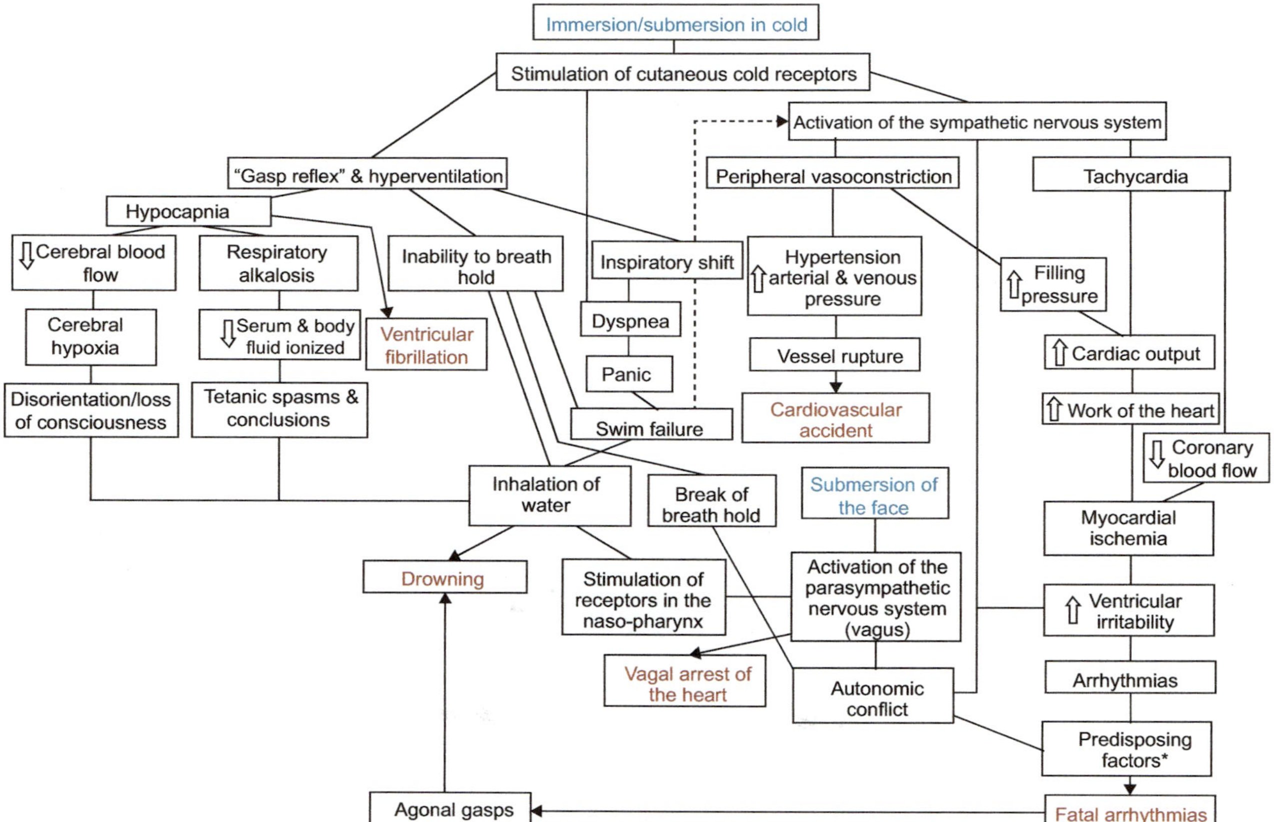

Figure 4.2 A contemporary view of the initial responses to immersion and submersion in cold water ("cold shock"). *Predisposing factors include channelopathies, atherosclerosis, long QT syndrome, myocardial hypertrophy, ischemic heart disease. *(Based on: Tipton (1989), Datta and Tipton (2006), Tipton et al. (2010), and Shattock and Tipton (2012).)*

United States in 1979 two HH-3F helicopters ditched in water temperatures of 13 and 14 °C. Of a total of nine crewmen, only three survived. None of those who perished had injuries extensive enough to have prevented their escape from the inverted, floating craft. The postcrash investigation revealed that all victims had drowned while attempting to egress. In each accident, decreased breath-hold time, owing to the effects of sudden immersion in cold water, was implicated as the likely precursor to drowning (Eberwein, 1985).

As mentioned, by the mid-1980s the preoccupation with hypothermia that had begun in the early years of the century remained. In comparison with the fatalities that occurred owing to hypothermia at the surface of the sea following egress from ditched inverted and floating helicopters, those occurring because of an inability to escape (cold shock leading to drowning or cardiac arrest) received relatively little attention. For example, in the United Kingdom, a large part of the comprehensive fatal accident inquiry into the *Cormorant Alpha* helicopter ditching of 1992, in which 11 of the 17 occupants died when their helicopter ditched next to an installation on its way to an accommodation platform, discusses the six fatalities that occurred at the surface of the sea following escape (Jessop, 1993). The quality of immersion suits, hypothermia, survival time, and factors influencing this time are considered. Those that failed to escape from the helicopter, in some cases despite having undone their seat belts, are described as "overcome by the sea." No consideration is given to why they were overcome, or what could have been provided with to help. Such a focus on hypothermia and lack of consideration of the "cause of the cause" of death of those failing to escape from ditched helicopters probably contributed to the delayed introduction of EUBA.

In 1995 it was estimated by groups such as the UK Coast Guard, military, and civilian operators in the North Sea that the time required for a controlled exit by all passengers in a ditched inverted helicopter is 40–60 s (Tipton et al., 1995). The short-fall between this time and the average maximum breath-hold time achieved on submersion in cold water wearing an uninsulated helicopter passenger suit (approximately 20 s) provided a powerful rationale for the provision of some form of EUBA. Indeed, a review of helicopter offshore safety and survival by the UK Civil Aviation Authority in the same year (CAA, 1995) provided details of four survivable accidents involving UK-operated offshore helicopters between 1976 and 1993. In these accidents 19 of 54 passengers died; 11 of those who died failed to escape from the helicopter, and eight died at the surface of the sea. It was recognized in the report that escape from a submerged helicopter may take longer than the time that a victim can be expected to hold his or her breath—especially if the water is cold. Despite this it was concluded that no clear advantage would be gained (by the provision of some form of underwater breathing device) and that, on the basis of evidence currently available, the CAA would not be justified in pursuing this as a regulatory measure.

Ten years earlier, in 1985, two major oil companies in the United Kingdom (Shell UK and Esso) had, in collaboration with the Royal Navy and University of Surrey, commissioned research into "submerged helicopter escape and survival." The resulting experimentation identified the initial response to immersion as a particular hazard for helicopter passengers and crew and one that was not ameliorated by either the "shorty" wet suits being used at the time to transport workers in the United Kingdom between installations and accommodation platforms, or the uninsulated immersion "dry" suits used to transport workers to and from shore (Tipton & Vincent, 1989). It was concluded that "the problems created by the inability of individuals to breath-hold during cold water submersion could, to some extent, be avoided by providing some form of emergency breathing system" (Brooks & Tipton, 2001).

In response to this recommendation, the oil companies challenged scientists, designers, and manufacturers to produce an EUBA that was simple in design and that, when used as recommended, can only be of assistance in significantly extending the underwater survival time of the user. This specification had some significant implications for design. In particular, the phrase "can only be of assistance" ruled out the use of sources of compressed air such as the HEED devices becoming available at the time. These either were mini SCUBA sets ("pony" bottles) or included cylinders of compressed air, which would introduce the potential danger of a pulmonary overpressure accident. Both spontaneous pneumothorax and arterial gas embolism have been reported on ascent from 1 m during training with helicopter EUBA (Benton in Benton, Woodfine, & Westwood, 1996; Tipton et al., 1995).

After detailed consideration of the literature relating to the control of breathing (e.g., Fowler, 1954), the concept of a simple rebreather was formulated and developed during the late 1980s and early 1990s (Tipton et al., 1995, 1997). This device became known as an "air pocket" and was manufactured by The Shark Group, the only company, of several that had been approached, who had agreed to participate in the original exercise.

In view of the long, and largely unresolved, history of incompatibility between immersion suits and life jackets (RGIT, 1988), it seemed important that if an EUBA was going to be added to the protective equipment provided for helicopter passengers (immersion suit and life jacket); the resulting ensemble should form an integrated survival system (ISS; Tipton, 1993). For the helicopter passenger this was to include advanced anti-hypothermia protection; a lifejacket, which, unlike many, would self-right a casualty wearing an immersion suit; and an EUBA (see Chapter 9). The fundamental principles behind this concept were that an immersion casualty should be provided with some protection against all of the hazardous responses associated with immersion in cold water, and that the individual components of an ISS should be compatible and complementary; they will also be interdependent in that the better the immersion suit the smaller the demand placed on the EUBA. As individuals are still being provided with different pieces of protective equipment that have been designed, developed, and evaluated separately, the ISS concept remains as applicable today as it was over two decades ago.

Finally, after about a decade, the provision of EUBA was added to the list of equipment and training provided to help people escape from ditched inverted helicopters, that is, immersion suits with air valves, underwater illumination of windows and exits, and training in inverted helicopter underwater escape.

In the intervening 20 years, many of the objections to the provision of compressed air have, in some quarters, been forgotten, ignored, or mitigated, and such devices are in widespread use in the industry (see Chapters 2 and 8). For safety reasons, close-to-surface training is employed with compressed-air EUBA, using a shallow water escape training (SWET) chair rather than full helicopter underwater escape training (HUET) in a "dunker," which requires egress from greater depths (>1 m). Outstanding issues with the provision of EUBA include:

- The difficulty associated with testing EUBA in a safe way that avoids the possibility of autonomic conflict (Figure 4.2).
- The, probably false, assumption that during an emergency, civilian helicopter ditching passengers will be able to deploy, activate, and use EUBA after immersion.
- The assumption in some quarters that "dry" training without immersion/submersion is sufficient or that near-surface training (SWET) is sufficient preparation for a helicopter ditching.
- The medical and technical provision required for the overpressure accidents that can occur with sources of compressed air during in-water

HUET, as opposed to SWET training. Between January 2001 and January 2006, a total of 15,000 Royal Navy (RN) trainees (3000 per annum) received STASS training. During this period, 34 RN trainees were referred to a medical center, including seven cases in which STASS, a source of compressed air, was being used (medical problem may not have been caused by STASS). The most serious case was a spontaneous pneumothorax.

- Guidelines (e.g., British Thoracic Society, 2003) on the fitness to dive suggest training with compressed-air diving should not be undertaken by those with a variety of chronic lung diseases, acute chest infections, and asthma. This could represent up to 10% of the population at any given time.

4.3 CONCLUSION

The initial responses to cold water immersion represent the greatest threat to be faced by those accidentally immersed in cold water, and particularly helicopter passengers who have to undertake an underwater escape. The personal protective equipment provided for such passengers should include an immersion dry suit with advanced anti-hypothermia protection, a life jacket, and an EUBA. The type of EUBA provided will be determined by an assessment of the value of the type of training that can be provided (HUET vs SWET) and the medical exclusions and risks associated with each type of device. The speed of deployment of a device underwater should be independent of whether it is a rebreather or source of compressed air.

When previously considered in the late 1980s, it was thought prudent to provide a simple rebreather; in the intervening years compressed air has gained in popularity, possibly because many of those associated with HUET are trained divers. Whatever type of device is selected, the various pieces of protective equipment provided for helicopter passengers should constitute an integrated survival system.

ACKNOWLEDGMENTS

Thanks to all those colleagues and volunteers who supported our work, especially Mike Vincent, David Elliott, Dave Stubbs, Neville Rendall, Pete Moncaster, Ian Botterill, and D. Litchfield. This chapter is dedicated to the memory of Eric Bramham, Frank Golden, and Tom Beames.

REFERENCES

Alexander, L. (1945). *The treatment of shock from prolonged exposure to cold, especially water.* London: Combined Intelligence Objectives Sub-Committee APO 413 C105. Item No. 24, Her Majesty's Stationary Office.

Benton, P. J., Woodfine, J. D., & Westwood, P. R. (1996). Arterial gas embolism following a 1 – meter ascent during helicopter escape training: a case report. *Aviation, Space, and Environmental Medicine, 67*, 63–64.

British Thoracic Society. (2003). *Guidelines on respiratory aspects of fitness for diving.*

Brooks, C. J., & Tipton, M. J. (2001). The requirements for an emergency breathing system (EBS) in over-water helicopter and fixed wing aircraft operations. In *RTO ARGAR-Dograph* (Vol. 341). France: Research and Technology Organization, North Atlantic Treaty Organization.

de Burgh Daly, M., & Angell-James, J. E. (1979). The 'diving response' and its possible clinical implications. *Internal Medicine, 1*, 12–19.

Civil Aviation Authority (CAA). (1995). *Review of helicopter offshore safety and survival.* London, UK: CAP 641. ISBN: 0-86039-608-8.

Cooper, K. E., Martin, S., & Riben, P. (1976). Respiratory and other responses in subjects immersed in cold water. *Journal of Applied Physiology, 40*, 903–910.

Currie, J. (1805). Medical Reports, on the effects of water, cold and warm, as a remedy in fever and other diseases, whether applied to the surface of the body, or used internally. In *Including an inquiry into the circumstances that render cold drink, or the cold bath, dangerous in health, to which are added; observations on the nature of fever; and on the effects of opium, alcohol, and inanition* (4th, Corrected and Enlarged ed., Vol. 1, p. ii). London: T. Cadell and W. Davies. Retrieved 2 December 2009. Full text at Internet Archive archive.org.

Datta, A., & Tipton, M. J. (2006). Respiratory responses to cold water immersion: neural pathways, interactions and clinical consequences. *Journal of Applied Physiology, 100*(6), 2057–2064. Review.

Eberwein, J. (July 1985). *The last gasp.* U.S. Naval Institute Proceedings, 128132.

Fowler, W. S. (1954). The breaking point of breath holding. *Journal of Applied Physiology, 6*, 539–545.

Golden, F. StC., & Hervey, G. R. (1981). The "afterdrop" and death after rescue from immersion in cold water. In J. A. Adam (Ed.), *Hypothermia ashore and afloat.* Aberdeen: Aberdeen University Press.

Golden, F., Hervey, G. R., & Tipton, M. J. (1994). Circum-rescue collapse: collapse, sometimes fatal, associated with rescue of immersion victims. *South Pacific Underwater Medicine Society Journal, 24*(3), 171–179.

Goode, R. C., Duffin, J., Miller, R., Romet, T. T., Chant, W., & Ackles, A. (1975). Sudden cold water immersion. *Respiration Physiology, 23*, 301–310.

Home Office Report. (1977). *Report of the working party on water safety.* London: HMSO.

Jessop, A. S. (1993). *Determination of the cormorant alpha fatal accident inquiry.* Aberdeen.

Keatinge, W. R., & Evans, M. (1961). The respiratory and cardiovascular response to immersion in cold and warm water. *Quarterly Journal of Experimental Physiology, 46*, 83–94.

Keatinge, W. R., McIlroy, M. B., & Goldfien, A. (1964). Cardiovascular responses to ice-cold showers. *Journal of Applied Physiology, 19*, 1145–1150.

Keatinge, W. R., & Nadel, J. A. (1965). Immediate respiratory response to sudden cooling of the skin. *Journal of Applied Physiology, 20*, 65–69.

Modell, J. H. (1971). *Pathophysiology and treatment of drowning.* Springfield, Illinois: Charles C Thomas.

Robert Gordon Institute of Technology (RGIT). (1988). *In-water performance assessments of lifejacket and immersion suit combinations.* London: RGIT Report for the Dept. of Energy, HMSO.

Shattock, M., & Tipton, M. J. (2012). "Autonomic conflict": a different way to die on immersion in cold water? *Journal of Physiology, 590*(14), 3219–3230.

Tipton, M. J. (1989). The initial responses to cold-water immersion in man. Editorial Review. *Clinical Science, 77*, 581–588.

Tipton, M. J. (1993). The concept of an "Integrated Survival System" for protection against the responses associated with immersion in cold water. *Journal of the Royal Naval Medical Service, 79*, 11–14.

Tipton, M. J., Balmi, P. J., Bramham, E., Maddern, T., & Elliott, D. H. (1995). A simple emergency underwater breathing aid for helicopter escape. *Aviation Space and Environmental Medicine, 66*, 206–211.

Tipton, M. J., Franks, C. M., Sage, B. A., & Redman, P. J. (1997). An examination of two emergency breathing aids for use during helicopter underwater escape. *Aviation, Space & Environmental Medicine, 68*(10), 906–913.

Tipton, M. J., Gibbs, P., Brooks, C., Roiz de Sa, D., & Reilly, T. (2010). ECG during helicopter underwater escape training. *Aviation, Space and Environmental Medicine, 81*, 399–404.

Tipton, M. J., Kelleher, P., & Golden, F. (1994). Supraventricular arrhythmias following breath-hold submersions in cold water. *Undersea and Hyperbaric Medicine, 21*(3), 305–313.

Tipton, M. J., McCormack, E., & Turner, C. (2014). An international data registration for accidental and immersion hypothermia: the UK National Immersion Incident Survey Revisited. In J. Bierens (Ed.), *Drowning, prevention, rescue, treatment.* Berlin, Heidelberg: Springer-Verlag.

Tipton, M. J., Stubbs, D. A., & Elliott, D. H. (1990). The effect of clothing on the initial responses to cold water immersion in man. *Journal of the Royal Naval Medical Service, 76*(2), 89–95.

Tipton, M. J., Stubbs, D. A., & Elliott, D. H. (1991). Human initial responses to immersion in cold water at 3 temperatures and following hyperventilation. *Journal of Applied Physiology, 70*(1), 317–322.

Tipton, M. J., & Vincent, M. J. (1989). Protection provided against the initial responses to cold immersion by a partial coverage wet suit. *Aviation Space and Environmental Medicine, 60*(8), 769–773.

CHAPTER 5

Psychophysiological Responses to Immersion: Preparation and Habituation

Martin Barwood
Department of Sport, Exercise and Rehabilitation, Northumbria University, Newcastle Upon Tyne, UK

5.1 BACKGROUND

The "will to survive" is a phrase often used to explain why, in exceptional cases, an individual surpasses the boundaries of human tolerance and physiological limitation in a life-threatening and sometimes hostile environment to achieve survival against the odds. The inference is that some aspect of a successful outcome in extreme cases of survival can be explained by an individual's psychological approach and his or her coping resource to deal with the demands of the situation (Barwood, 2005). In the experimental setting it is difficult to assess this concept, as it is highly problematic and ethically questionable to attempt to simulate a truly life-threatening scenario. An alternative is to retrospectively explain successful and unsuccessful outcomes from survival scenarios, based on the available evidence, hoping to reach a consensus on the likely influential variables that determined the outcome. Unfortunately evidence in such cases is often absent, highly anecdotal, or unreliable, making a stable conclusion difficult to reach. Consequently, evidence-based interventions drawn from this type of retrospective observation may prove to be questionable, unsafe, and without adequate foundation. In the experimental setting under controlled conditions, we can simulate the environmental stressors, but to a lesser magnitude, to observe or manipulate the physiological and psychological variables that are known to influence the outcomes of accidental, life-threatening exposures in the real world. We are assuming that the relationships we observe in the laboratory also hold true at the extremes when a threat to life is sensed. Chapter 4 (Tipton) considered some of the physiological factors that dictate survival of accidental cold-water immersion or submersion and helicopter underwater escape. The present chapter

Handbook of Offshore Helicopter Transport Safety
ISBN 978-1-78242-187-0
http://dx.doi.org/10.1016/B978-1-78242-187-0.00005-5

considers the evidence that psychological aspects and approaches could also play a role in such situations. It also considers what can be done by way of preparation for an individual at risk of cold-water immersion.

5.2 INTRODUCTION

As is described in Chapter 4, accidental immersion in cold water is a life-threatening situation because of the rate of onset and magnitude of the cold shock response, thereby increasing the risk of aspirating water to the lungs and drowning. Consequently, basic physiological functions such as breathing at a normal frequency and maintaining a normal heart rate are nearly impossible, at least for the first few minutes of immersion. The accidental immersion scenario also elicits a range of psychological responses that can either exaggerate (e.g., acute anxiety) or enable the suppression of (e.g., pre-practiced psychological training strategies) components of the cold shock response, thereby influencing the chances of survival. If an individual is at daily risk of immersion, as in the case of the aviator and aircrew, it is wise to consider the extent to which psychological variables can also influence the chances of survival. Where possible, practitioners should consider incorporating psychological strategies into their survival training programs that have been shown to be influential and beneficial in such circumstances. If the immersion scenario includes helicopter ditching and a requirement to egress an inverted craft, it is likely that the psychological demands of the situation will compound the physiological responses that arise. The more demanding the survival scenario, the more it can be expected that psychological factors and strategies could play a significant role in those who are accidentally exposed. This could potentially increase their chances of becoming a victim rather than a survivor (Leach, 1994).

5.3 PREPARATION FOR COLD-WATER IMMERSION: PSYCHOLOGICAL INFLUENCES ON THE PHYSIOLOGICAL RESPONSES

In very few cases do individuals have adequate warning time to prepare for an accidental cold-water immersion, hence the rationale for survival training for those at daily risk of cold water entry. The assumption is that the techniques learned through practice in controlled simulation conditions are employed automatically when the emergency scenario is faced. We employed the same principle when psychologically preparing individuals for cold-water immersion in the laboratory scenario and the results were both remarkable and highly beneficial in influencing the respiratory component of the cold shock response (Barwood, Dalzell, Datta, Thelwell, & Tipton, 2006). We chose to study breath-hold time on immersion, as it was our best index of the ability to resist the ventilatory drive elicited by sudden skin cooling that drives the cold shock response. Logically, the longer a breath-hold could be maintained in water, the lower the chances were that water could enter the lungs and precipitate drowning. It is commonly accepted that maximal breath-hold time on immersion is a fraction of that which can be achieved in air, and experimental data suggest that there is probably a relationship between the two (Tipton, 1992). At the extreme, the breath-hold time on cold-water immersion can be reduced to only fractions of a second (i.e., 0.2 s) in unprotected individuals in very cold water. Tipton and Vincent (1989) observed this extreme and reported that the average breath-hold time fell from 45 to 9.5 s when participants were immersed in 5 °C water compared to breath holding in air. We saw an 80% improvement in maximal breath-hold time on immersion with the participants who received a psychological intervention strategy, improving from an average of 24.7 s prior to psychological intervention to 44.3 s postintervention, nearly a 20 s improvement. This contrasted to no improvement in the initial breath-hold time on immersion in a matched control group (initial breath-hold time of 24.0 s), who recorded a slightly lower breath-hold time on the second immersion (21.3 s; 2.7 s lower). In this experiment we were careful to match the water temperatures (12 °C) that the participants were exposed to (within participant) between immersions 1 and 2 to within 0.1 °C. This control of water temperature allowed us to be confident that the thermal drive from the cold water was the same. From the results of the study we could conclude that we improved the capability to tolerate and suppress the drive to breathe despite the same magnitude of cold shock response. If we consider that a minimum breath-hold time of 28 s on immersion is required to egress a ditched and inverted helicopter under ideal conditions (Cheung, D'Eon, & Brooks, 2001), it seems possible that the psychological component manipulated by this intervention was sufficient for individuals to cross the boundary of being a "victim" (i.e., insufficient initial breath-hold time of 24 s to enable helicopter egress) and increase the chances of "survival" (i.e., exceeding the minimum threshold for helicopter egress). Clearly, it is difficult to uphold the suggestion these data make about the importance of psychological aspects in a real-life emergency scenario, yet it is compelling evidence that preparatory psychological strategies can help suppress components of the cold shock response, which is known to be life-threatening (Tipton, 2003).

5.4 VARIATION IN THE RESPONSE TO PSYCHOLOGICAL INTERVENTION

Despite the promise of the beneficial effects of these preparatory strategies, the inter-individual variability in response to psychological intervention was still large. Indeed, the range in postintervention maximal breath-hold times was between 13.5 and 115.9 s, with the range before intervention being between 5.0 and 61.7 s (Barwood et al., 2006). We noted that it was the individuals who had the longest initial breath-hold time in air and cold water who tended to improve to the greatest extent as a consequence of the psychological intervention. Yet, we could find no common thread from the data we collected to predict this relationship as it was not consistent across participants. These data suggest that the relationship between the ventilatory component of the cold shock response and the psychological variables is complicated, but we are more likely to be able to influence the physiological response over which there is greatest voluntary control. For example, within this study and others preceding it (see Barwood, 2005) we examined whether individuals could exert voluntary influence over physiological responses such as heart rate and blood flow through structured psychological interventions. The suggestion is that we could gain voluntary control over responses that were otherwise thought to be automatic. We found no evidence to support this idea, as changes in heart rate and blood flow in the immersion scenario are driven by strong afferent signals through the autonomic nervous system, which is under less voluntary influence. Even in advance of the immersion, the pre-immersion heart rate response and tension rating (how tense they felt about the impending immersion) were unchanged (Barwood et al., 2006). In contrast, ventilation (i.e., breathing) can be placed under a relatively high proportion of voluntary control, the epitome of which is breath-holding.

5.5 PSYCHOLOGICAL COMPONENTS TO BREATH-HOLDING IN AIR AND WATER

In air, the drive to break a breath-hold is dictated by two factors. The first is the capability to withstand the buildup of afferent feedback from static lung stretch receptors that would normally be active during a standard respiratory cycle (Godfrey & Campbell, 1968). The consequence of these stretch receptors being static rather than active is a buildup of efferent drive from the respiratory center of the brain to the respiratory muscles to recommence breathing movement (Godfrey & Campbell, 1968).

Relieving this drive without refreshing the oxygen content of the lung explains why a rebreather system proves an effective underwater breathing apparatus in the short term (Tipton, Franks, Sage, & Redman, 1997). That is, the mechanics of breathing are not initially driven by the increasingly low oxygen environment (i.e., an hypoxic environment) and consequent buildup of carbon dioxide (i.e., hypercapnia) from ongoing cellular respiration (see Datta & Tipton, 2006, for a detailed review of the respiratory responses to cold water). In the scenario in which a rebreather is used, the development of an hypoxic and hypercapnic environment would be the secondary (from a temporal perspective) drive to break the breath-hold (Tipton, Balmi, Bramham, Maddern, & Elliot, 1995). At this stage, the noxious sensation of air hunger would start to increase, and breaking of the breath-hold may become inevitable. Both of these factors are subject to the limitations of an individual's psychological tolerance of the noxious sensations that arise. On breath-holding during entry into cold water, the hyperventilatory drive comes initially from an increase in sympathetic nervous system input to the respiratory centers of the brain as a consequence of sudden skin cooling (Keatinge & Evans, 1961). This represents a new phase of the respiratory drive to withstand specific to cold-water immersion. It is unlikely that any meaningful drive from static stretch receptors would arise until later in the breath-hold. Hence, we think that the psychological demands on breath-holding in cold water are slightly different from those required in air. However, an ability to psychologically withstand the noxious sensations associated with each phase will enable longer breath-holds to be achieved. It may be that those individuals who are best able to tolerate these sensations in air would also be the ones who stood the best chance, on balance, of maintaining a breath-hold in cold water.

Breath-holding is not without its dangers, however. Indeed, Tipton, Gibbs, Brooks, Roiz de Sa, and Reilly (2010) noted that coincidental stimulation of both divisions of the autonomic nervous system during submersion (i.e., the sympathetic nervous system—the excitatory division—and the parasympathetic nervous system—the inhibitory division), particularly at the break of the breath-hold, had a tendency to produce abnormal heart rhythms (arrhythmias), thereby inducing a further threat to life in persons with underlying cardiovascular disorders (for a full description of this phenomenon, see Tipton et al. (2010) and Chapter 4). Hence, our advice in cases of accidental immersion (i.e., head out of the water), rather than submersion (head under, as may be the case with helicopter escape), is to "float first" where possible until the cold shock subsides, following which an exit survival strategy can be sought (Barwood, Bates, Long, & Tipton, 2011).

5.6 ANXIETY AND THE COLD SHOCK RESPONSE

In the emergency scenario it is logical to suggest that negative emotional states will arise that could compound the magnitude of the physiological responses (Barwood et al., 2012). For example, if preparatory psychological intervention strategies are beneficial to controlling the cold shock response, then it is also possible that negative emotional states such as acute anxiety experienced in advance or during an immersion could also increase the magnitude of the response and lessen the chances of surviving accidental immersion. We investigated this possibility, but chose not to include breath-holding in that study because of the impending risk of arrhythmias (Tipton et al., 2010). We uncovered an interesting relationship between anxiety and cold shock. In this experiment participants underwent two 7-min cold water (15 °C) immersions in a counterbalanced order. Prior to one of the immersions, the participants were told that the water was going to be 5 °C colder (i.e., 10 °C) than they were expecting; in fact, it remained unchanged. The effect of this suggestion was a significant increase in the pre-immersion anxiety rating that resulted in significantly elevated heart rate before and throughout the 7-min immersion. The extent of the influence of the increase in anxiety was also sufficient to increase the ventilatory response before and during the immersion, although more so toward the latter part of the 7-min period. These data suggest that acute anxiety actually increases the extent of the cold shock response in persons who have not been repeatedly exposed to cold water (i.e., they were not habituated), and the extent of the influence is large enough to extend to the responses over which we have greatest voluntary control (i.e., breathing). It could be that in previous experiments using psychological interventions (i.e., Barwood et al., 2006) the amount of anxiety experienced by some individuals, irrespective of whether they had received a psychological intervention or not, was sufficient to override their psychological capabilities to suppress the drive to breathe, and a poor breath-hold performance resulted; this may account for some of the evident inter-individual variation. In the real-life emergency scenario we can only assume that acute anxiety is high and the consequent effects on the cold shock response are to increase the cardiac and respiratory components to a near-maximal extent (depending on water temperature).

Controlling the associated anxiety and lessening this perceptual component of the cold shock response may be one mechanism by which psychological strategies and/or survival strategies serve to benefit persons who are accidentally immersed.

5.7 PREPARATORY PSYCHOLOGICAL STRATEGIES AND COLD-WATER IMMERSION

It is important to describe the psychological strategies that have proven to be effective in improving maximal breath-hold time on immersion in the experimental setting in order that they could be incorporated into the survival training programmes and situations that include cold water as a stressor. The rationale for their inclusion in the original experiments was borrowed from the sports science literature in which psychological skills training (PST) is an established method for significantly improving sports performance (Thelwell & Maynard, 2003). The theory is that skills that have been shown to enhance performance in the most demanding of sporting situations could be tailored to positively influence the responses to cold-water immersion. PST often comprises a package of skills that are tailored specifically to meet the demands of a situation or sport. In theory, the components of PST that were included in the original study worked by focusing attention (goal-setting), modifying physiological arousal (arousal regulating strategies in the form of relaxation), enabling mental rehearsal (mental imagery), and improving self-confidence (positive self-talk). The collective qualities of PST packages were well supported at the time of use in the first cold-water immersion study. Yet, the shortcoming of the approach of using a package of skills is that it is unclear as to what contribution each skill makes to the experimental effects that are seen. Moreover, little is known about the neurophysiological processes that underpin PST and it is possible that a large part of the effect is simply due to a placebo. Despite these limitations, sufficient evidence does exist from the sports setting to substantiate the likely efficacy of PST in other situations, as we subsequently demonstrated (Barwood et al., 2006; Barwood, Datta, Thelwell, & Tipton, 2007; Barwood, Thelwell, & Tipton, 2008). A brief outline of each skill and its theoretical relationship with the cold shock response is included below.

5.7.1 Goal-Setting

It is difficult to imagine that in a survival situation, for example, a helicopter ditching and the consequent requirement to perform underwater escape, an individual may require a goal or target—to preserve life being the obvious one. Yet, in such situations the ability to focus on relevant cues, for example, timing the onset of breath-holding, navigating through an upturned craft, or deploying an emergency underwater breathing aid, while filtering out irrelevant information such as loud noise or passing debris may be fundamental to the success of the survival attempt. This is particularly relevant since it is likely that, under such extreme duress, a limited capability to focus on relevant cues in the form of perceptual narrowing may arise (Leach, 2005). Therefore, the ability to focus attention is limited at a time when optimal performance is paramount. Leach (2005) describes the possibility of a malfunction of the attentional processing system (specifically the supervisory attentional system), the consequences of which could prove to be fatal in a survival situation for those who are untrained. In our studies on cold-water immersion while breath-holding (Barwood et al., 2006, 2007), the outcome goal (i.e., an extended breath-hold time equating to improved chances of survival) was assumed and we instead focused on the processes that would culminate in achieving the outcome. For example, participants were required to identify the points at which the drive to break the breath-hold was greatest (e.g., when the water reached the torso) and how they would deploy their efforts to overcome this urge (e.g., use of their PST techniques). The process of focusing on the constituent parts of a task such as helicopter underwater escape through a structured, goal-focused survival training program may help a potential victim function instinctively in an emergency situation. The goal is to minimize the attentional processing requirements of the task and consequently reduce the chances that the demands of the situation become overwhelming.

5.7.2 Arousal Regulation

A discrepancy between the perceived requirements of a situation and the coping resources to deal with it can culminate in the negative emotional state of anxiety and that can be expressed, in part, by an increase in physiological arousal (Hackfort & Schwenkmezger, 1993). We have presented evidence that acute anxiety experienced prior to and during cold-water immersion has negative physiological consequences in the form of increased peak and average heart rate and breathing frequency. Collectively, these may represent an increase in physiological strain experienced on immersion and heighten the risk of aspirating water to the lung and drowning (Barwood et al., 2012). It would obviously be advantageous to reduce the extent of these increases in order to lessen the effects of cold shock. Arousal regulation strategies were tailored to combat the physiological consequences of anxiety by introducing participants to techniques that would induce relaxation (i.e., progressive muscular relaxation techniques) and control breathing (i.e., centering) prior to the second cold-water immersion (i.e., after PST) (Barwood et al., 2006). In a real-life scenario it is unlikely that these techniques could be practiced in full, given the time constraints, and we saw no obvious physiological consequence of their effects by way of a lower anticipatory heart rate prior to immersion. However, participants still rated their inclusion in the PST package as useful.

5.7.3 Mental Imagery

Visualization techniques in the form of mental imagery would be unlikely to be of direct use in the emergency situation. Rather, they could be tailored to enable mental rehearsal of the temporal patterning of the procedures advanced through survival training and/or the deployment of PST; in the experiments examining breath-holding capability upon cold-water immersion we used it for the latter. For example, it might not be practically and economically feasible or a statutory requirement to undergo survival simulation training on a frequent enough basis to enable unimpaired recall of the processes to complete helicopter underwater escape. Consequently, it may be of use to employ mental recall and visualization of the sequence of events that comprise successful escape of a ditched and inverted helicopter as a support skill; preflight and in-flight safety videos clearly provide important information that could support this technique. Also, the participants in our original experiments reported that, after their first immersion, the cold water was more disabling of their responses than they expected. Consequently, they felt more apprehensive of their impending second immersion. It was therefore imperative that they visualized themselves performing their second breath-hold immersion (i.e., after PST) in a different and more successful way. In the context of these experiments, visualization and mental imagery were used to consolidate the new processes to be implemented after PST into one clear program of actions to be deployed sequentially as the immersion ensued. Participants were conditioned to see themselves performing successfully and consequently reduce the apprehension (anxiety) associated with immersion, thereby improving their self-confidence.

5.7.4 Positive Self-Talk

We can only speculate that persons placed in a life-threatening circumstance, such as helicopter ditching, experience negative cognitions in conjunction with the physical sensations. We know that participants undergoing cold-water immersion often experience negative cognitions in conjunction with the event. For example, when immersed in 11 °C water, participants have reported that it was "too cold" and that they felt "I'll never hold my breath" (Barwood et al., 2006). In the sporting arena, negative cognitions are associated with distraction from key task-related cues and are synonymous with cognitive anxiety, with the somatic component of anxiety synonymous with an unwanted increase in physiological arousal (Conroy & Metzler, 2004). Consequently, unwanted negative thoughts have the potential to reduce self-confidence and increase the chances of poor task performance. Positive self-talk statements were tailored for the demands of cold-water immersion to guide the participants toward relevant, task-related cues that triggers for deployment of their PST techniques. For example, participants identified the point at which the water reached their torso as most demanding and focused upon "getting through the initial shock of immersion to maintain the breath-hold" (Barwood et al., 2006). In advance of the second cold-water immersion, participants were asked to identify the negative cognitions they experienced in the first immersion and recognize the consequent effects of those experiences. They then developed positive self-talk statements to replace those negative cognitions and to guide themselves toward the relevant processes for a successful breath-hold performance on the final cold-water immersion. This part of the intervention was designed to be complementary to the content of the goal-setting component of the PST package, the components of which are summarized in Figure 5.1.

5.8 EXPERIMENTAL EVIDENCE FOR THE EFFICACY OF PST DURING COLD-WATER IMMERSION

After the completion of each study using PST (Barwood et al., 2006, 2007) participants were asked to rate how useful they found each component of the package on an 11-point Likert scale prior to and during cold-water immersion; the scale ranged from 0 (not at all useful) to 10 (very useful). There was no clear difference between the ratings of each of the components of the PST package, supporting the efficacy of the collective use; all skills were rated between 6 and 9 on the Likert scale. The PST tended to be rated more highly for use during the immersion (8/10) rather than in preparation (7/10), with the highest rating evident for positive self-talk use during immersion (9/10; Barwood et al., 2006) and the lowest evident for

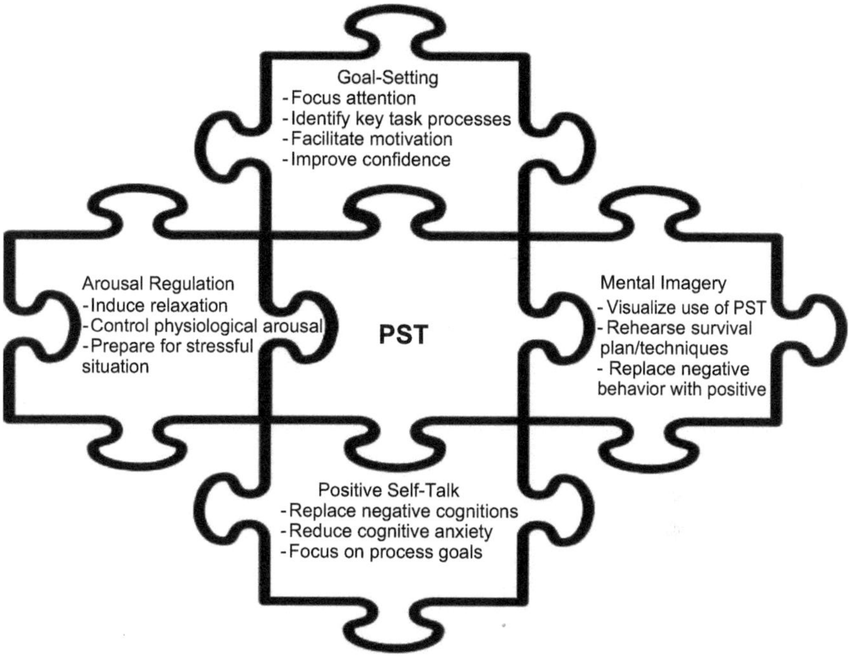

Figure 5.1 Summary of the components of PST and their rationale for use prior to and during immersion.

mental imagery use prior to immersion (6/10). Collectively, all skills were rated above the neutral rating on the scale (i.e., above 5/10).

5.9 OTHER MEANS OF PSYCHOLOGICAL PREPARATION

PST is one means of psychological preparation that has been shown to be effective in the sporting domain and during laboratory-administered environmental stress in the form of cold water exposure (Barwood et al., 2006, 2007) and during exercise in the heat (Barwood et al., 2008). That is not to exclude other means of psychological preparatory strategy. For example, stress inoculation training (SIT) is another form of cognitive behavioral therapy primarily aimed at cognitive stress management in relation to a specific event (Meichenbaum, 1996).

SIT has been administered to aid individuals in dealing with anger (Levendusky & Pankratz, 1975), with chronic or experimentally induced pain (Ross & Berger, 1996) and to help occupational groups under regular daily stress (Sarason, Johnson, Berberich, & Siegel, 1979). Similar to PST, SIT is delivered in phases comprising education, rehearsal of the skills, application to the given situation, and positive thinking (Meichenbaum, 1996). It also encompasses some aspect of motivation toward overcoming a given stressor or situation, arousal-regulating strategies, mental imagery (primarily for rehearsal), and self-instructional training replacing negative cognitions with positive and instructional phrases similar to self-talk. The theoretical underpinning of SIT is more developed than that of PST and is based upon the rationale that conditioned reactions are self-activated on the basis of learned expectations associated with a given stimulus, rather than automatically evoked behaviors (Bandura, 1977). SIT is thought to aid the individual in reacting to the occurrence of an adverse event and, following training, substituting in an adaptive response to that stimulus. Both PST and SIT aim to cognitively remodel the initial negative behavior in response to a given stressor into a controlled and targeted positive response to the situation. The fundamental idea is that cognitive events refer to conscious and identifiable thoughts and images and that these events form an internal dialogue that takes place when an adverse event occurs. Both PST and SIT aim to alter this dialogue and therefore manipulate an individual's resistance to a given situation. In short, they aim to raise conscious awareness of the components underpinning a behavioral response to a stressor. Repetition of exposure to the stressor with the remodeled behavior may then reinforce the development of a new, adaptive response.

Experimental evidence for the efficacy of SIT is advancing and it is possible that SIT could also be tailored to help cope with life-threatening situations such as helicopter underwater escape. Of those studies utilizing SIT as the sole treatment, the efficacy of imagery rehearsal and role-playing was demonstrated in tolerating pain induced by submersion of the non-dominant hand into 2 °C water (Turk, 1975). Participants imagining the stressful situation integrated with images of themselves successfully using the SIT coping technique maintained the immersion for a significantly longer period. These findings were also demonstrated by Meichenbaum (1972) in subjects tolerating ischemic pain induced by the overinflation of a blood pressure cuff; pain was tolerated for an average of 17 min longer in SIT subjects. Likewise, Hackett and Horan (1980) reported a similar experimental effect with a cold pressor immersion task tolerated for a longer time in participants who underwent SIT.

Evidence of SIT for submersion during helicopter underwater escape can also be found. Brooks, Gibbs, Jenkins, and McLeod (2007) successfully remodeled an initially maladaptive behavior, and the resultant experience of acute anxiety, associated with escape training in one military pilot.

The pilot had developed a phobia of submersion and being trapped underwater in a previous training exercise as part of his military duties. Both submersion and being trapped underwater occur, at least temporarily, during helicopter underwater escape. Brooks et al. (2007) aimed to treat the pilot with a desensitization program that exposed the pilot in a step-wise manner to graded (by difficulty) water immersion ranging from simple vertical submersions up to the neck in water to complex inverted underwater cross-cabin escapes in full aircrew equipment. This treatment series relieved the anxiety associated with submersion and alleviated the phobia to an extent that the pilot was able to proceed with a standard helicopter underwater escape training course, enabling continuation of aircrew training. Brooks et al. (2007) based their intervention on an adapted version of SIT implemented by Banken and Mahone (1991), who used SIT to relieve an undergraduate student's experience of stress associated with pilot training. Brooks et al. (2007) focused more so on the practical components of the situation and did not deliver cognitive strategies per se; it is an assumption supported by the experimental outcomes that Brooks et al. (2007) showed that the cognitive appraisal of the submersion situation was changed as a consequence of the intervention the pilot received. Collectively it seems that either PST or SIT could be implemented to help with the demands of water immersion and submersion. More studies are required using either intervention to discern the efficacy and adaptability of PST and SIT to life-threatening scenarios.

5.10 HABITUATION TO COLD WATER: A PSYCHOPHYSIOLOGICAL RESPONSE?

Thus far we have presented the evidence that the responses evoked by cold-water immersion can be modified, in part, by psychological intervention. It is also possible to lessen the extent of the physiological response to cold by repeatedly exposing an individual to short-term cold-water immersion, thereby inducing an habituation (Golden & Tipton, 1988). Habituation is defined as a diminished response to a repeated stimulus of the same magnitude (Zald, 2003). Tipton, Mekjavic, and Eglin (2000) demonstrated that an habituation of the cold shock response occurs as a consequence of as few as five repeated cold-water immersions in 15 °C water of 3-min duration. Results indicated that the response was reduced by up to 40% of that evident on the first immersion. This blunting of the cold chock response was still statistically evident 7 months after the initial series of five immersions. Further tests were conducted after 14 months but only the cardiac component of the cold shock response remained significantly reduced, with the respiratory response having returned to the level seen on initial cold-water immersion 14 months prior. These responses were all contrasted to a control group that was not exposed to cold water in the intervening months between tests and who showed no evidence of an habituation in the first 30 s of immersion. Yet, the control group did demonstrate lower respiratory responses over the last 150 s of cold-water immersion despite not having been repeatedly exposed; this infers a perceptual habituation. More recently, the number of immersions required to blunt the cold shock response (statistically) was reduced from five to four (Barwood et al., 2007), thereby reducing the logistical demand of implementing an habituation regimen.

From a practical perspective, repeated whole-body cold-water immersions on consecutive days are straightforward to implement (e.g., an ordinary bath could be used) and the benefits of a reduced cold shock response can be seen for a period exceeding one year (Tipton et al., 2000). It is not clear what the maximum period of time is that can elapse between habituation immersions before no blunting of the cold shock response occurs; no experiments have been conducted to answer this important practical question. The water temperature in which a habituation regime should take place is a matter of some debate. For example, some authors feel it is unnecessary to complete a habituation in water temperatures below 10 °C because of the increased likelihood that pain receptors, in addition to thermoreceptors, will be activated; this is the ethical basis for the majority of recent work being conducted at or above 10 °C. Others feel that a lower limit of 5 °C is acceptable (Goode et al., 1975). It is clear that habituation achieved by serial immersion in water at 15 °C can significantly blunt the response seen at 10 °C (Tipton et al., 2000), providing a further rationale for completing habituation at higher (although still cold) water temperatures.

5.11 HABITUATION: A POSSIBLE MECHANISM

The mechanism by which habituation works is not entirely clear, yet experiments have revealed that it is a phenomenon specific to the central nervous system rather than a change in the thermal information that is sensed by the skin (Tipton, Eglin, & Golden, 1998). In these experiments, Tipton et al. (1998) first immersed all participants into cold water (10 °C) while exposing only the right-hand side of the body. The participants were then separated into a habituation ($n = 8$) and a control group ($n = 7$). The habituation group then underwent six further cold-water immersions with only the left-hand side of the body exposed, while the control group experienced no further cold water exposure before completing a final cold-water immersion in which only the right-hand side of the body was immersed. The resultant effect was no change in the cold shock response in the control group, whereas the habituation group showed a significantly lower cold shock response even though opposing sides of the body had been tested in the experimental immersions (left side of the body) compared to the habituation immersions (right side only). It seems that, irrespective of where the thermal information was derived from on the skin, the central nervous system modulated that information and elicited the cold shock response in accordance with the present and previous cold exposure. From a practical perspective, it seems that at least some whole-body protection against the cold shock response can be achieved (i.e., the response is blunted) by exposing only part of the skin to cold water. At this stage it is unclear, however, whether the rate of habituation that is achieved is accelerated or reduced in accordance with the skin surface area exposed to cold. It is clear that after approximately five immersions there is probably a diminishing return in the blunting of the cold shock response that takes place (i.e., the reduction in the cold shock response is not accelerated greatly) with each additional immersion that is conducted (Barwood et al., 2007).

Habituation can really be considered as an exclusively physiological response only as a convenient thought; yet it probably comprises both physiological and psychological aspects (Barwood, Corbett, & Wagstaff, 2014). A 2014 experiment showed that simply undergoing repeated *warm-water* immersion (35 °C) lessened the cold shock response seen in a second *cold-water* immersion (Barwood et al., 2014). This effect was reserved to the respiratory portion of the cold shock response. Specifically, the volume of air breathed in and out per breath (tidal volume) was marginally lower by about 150 mL per breath after having repeatedly experienced the immersion scenario in the absence of a repeated cold stimulus; hence we suggest a perceptual habituation took place that lessened the physiological response that was seen on cold-water immersion. Although this difference is relatively small it may be critical when you consider that the lethal volume of sea water that is needed to irreversibly damage gas transport in the lung is approximately 22 mL kg^{-1} (Modell, Gaub, Moya, Vestal, & Swarz, 1966), for example, 1.65 L in a 75-kg individual. Lower tidal ventilation would at least delay the point at which this threshold was reached and give more chance for the immersed victim to defend his or her airway. Likewise, the timing of the lower ventilation is also important given that tidal volume was seen to be numerically lower throughout the second cold-water immersion and statistically so from the second minute of the test. The cold shock response is perhaps most threatening in the early minutes of immersion (Tipton, 1989), following which the respiratory drive subsides and an exit strategy can be conceived. In hypothesizing a mechanism for this change in physiological response, we have suggested that repeatedly experiencing the immersion scenario lessens the primary or secondary appraisal of threat (Lazarus & Folkman, 1984) posed by the immersion scenario, resulting in a lower physiological response. This may be one beneficial means by which survival simulation training improves the prospect of survival in the emergency scenario. This experiment certainly does not suggest that the extent of physiological habituation that takes place with warm water immersion could approach that seen with repeated cold water exposure. It simply highlights that some of the benefit seen with repeated cold water exposure could be attributed to a perceptual habituation. It may be that a reduced appraisal of threat (i.e., perceptual habituation) associated with the immersion scenario is already integral to survival training. Indeed, Tipton et al. (2010) showed that the heart rate response to helicopter underwater escape training was lower after five "runs" (simulations) than after two despite only a modest thermal stimulus (water temperature of 29 °C) that is unlikely to induce a cold shock response per se.

5.12 THE COMBINED EFFECTS OF HABITUATION AND PST: A SYNERGISTIC EFFECT?

Given that repeated cold water exposure induces a physiologically relevant reduction in the cold shock response (Golden & Tipton, 1988), it was logical to explore the additive effect of habituation with PST on the breath-holding capability of immersed persons (Barwood et al., 2007). In this experiment participants first underwent an initial cold-water immersion during which they were required to breath-hold. They were then matched and allocated to two equal groups ($n = 10$) who both underwent repeated cold-water immersions (12 °C; on separate days) without breath-holding in sufficient number to blunt the cold shock response (i.e., five immersions). In addition to this, one of the groups also underwent a PST intervention similar to the one used previously (Barwood et al., 2006). The effects were that a 120% improvement in breath-hold time on immersion was seen compared to their first immersion; this group improved significantly from an initial breath-hold of 22.4 s to 49.3 s after receiving the intervention. By comparison, the improvement in the habituation-alone group was numerically lower than that of the other group but significantly higher than that seen on their first cold-water immersion with breath-holding. The habituation-only group improved from 22.0 to 36.1 s (i.e., a 73% improvement) as a consequence of repeated cold water exposure only. It is difficult to stop participants who undergo repeated cold-water immersion from developing their own psychological strategies to help cope with the situation. Yet, it seems that data from this experiment suggest that structured PST packages, similar to those we used in previous studies, are superior in influencing the cold shock response to those that are self-conceived. For convenience (a direct comparison across studies is not feasible), if we pool the magnitude of differences seen in breath-holding after three different interventions across the two studies (Barwood et al., 2006, 2007; see Figure 5.2), it is evident that no improvement in breath-holding is likely without intervention (i.e., the control group), whereas a

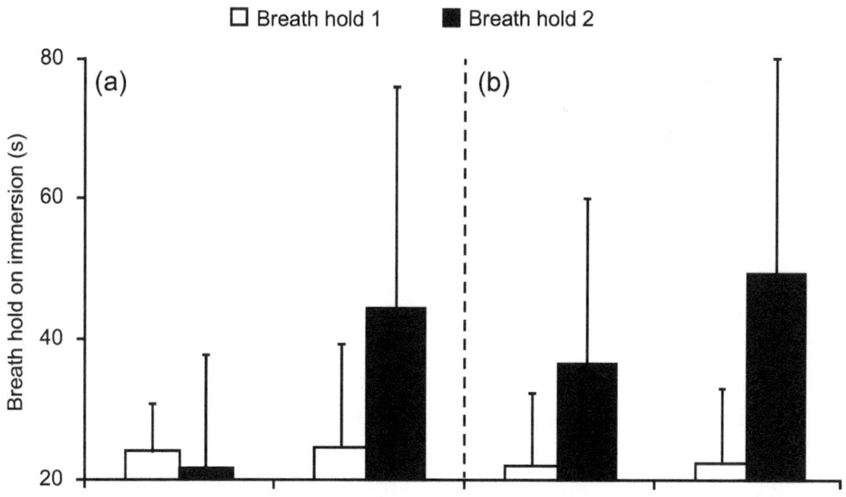

Figure 5.2 Improvements in breath-hold time that can be seen with no intervention (control) and following PST (PST), habituation (HAB), and habituation combined with PST (H-PST) from two separate studies (Barwood et al., 2006 (a). Barwood et al., 2007 (b)). A direct statistical comparison cannot be drawn between the two studies because of a slight difference in water temperature (11 °C vs 12 °C).

73% improvement might be seen with habituation alone. If we compare that to either intervention that included PST (i.e., an 80% improvement with PST alone and a 120% improvement with habitation combined with PST), then it is evident that the psychological component of the cold shock response is significant and meaningful and should be addressed within survival training where possible. Yet, it is also evident from these data that the variation in the response to PST with or without habituation is large by comparison to interventions that do not include PST (see Fig. 5.2.).

There is one further study that has utilized PST in this context but under more ecologically valid conditions. Croft et al. (2013) completed a habituation regimen combined with PST in inexperienced swimmers ($n = 6$), in conjunction with treading water, on rapid immersion; previous studies have used graded immersion. They found that the habituation combined with PST enabled improved control of the respiratory portion of the cold shock response and possibly lessened the reduction in brain blood flow that is seen ordinarily with cold shock; this reduction in blood flow is driven by low carbon dioxide levels that result from the uncontrollable hyperventilation induced by cold-water immersion. From a practical perspective this study confirms the role that PST can play in potentially improving survival prospects on cold-water immersion.

In contrast to PST, a partial reversal of habituation was evident when anxiety was induced after participants had completed a habituation regimen in an experimental condition cold-water immersion. In this study, anxiety was induced in the same way as that described earlier (i.e., by deception about the water temperature) and the resultant effect on the cold shock response was measured in participants who were free breathing on immersion. When participants were made to feel more anxious about the impending immersion, the cardiac component of the cold shock response was restored to the same level as that seen prior to habituation. That is, the coexistence of anxiety eradicated the change in heart rate response that was induced as a consequence of habituation; this was the basis for concluding a reversal of habituation had taken place. There was no effect on the respiratory portion of the response. We were the first to demonstrate this effect during whole-body immersion, but not the first to suggest that anxiety may have a modulating effect on the responses to cold water. Indeed, Glaser, Hall, and Whittow (1959) demonstrated a reversal of habituation of the cold pressor test when hand immersion was performed in front of a live audience thereby inducing anxiety in the participant. They concluded that afferent thermal and perceptual (affective) sensory information that arrived for integration within the central nervous system at the same time may interfere with the habituation response that was previously seen. We advanced this discussion and have speculated that the neural mechanism that is involved in this change is triggered by a change in the appraisal of threat posed by the impending cold(er) water immersion, culminating in a greater input to the hypothalamus from the limbic system (possibly the amygdala; Barwood et al., 2012). However, we have yet to test this idea. It is certainly now of interest to examine whether habituation is coincidental to a reduction in anxiety associated with immersion, as the reduction in the cold shock seen with habituation shares a similar time line with the lowering of anxiety associated with immersion, suggesting the two could be linked.

5.13 SUMMARY

It is evident that much can be done to prepare for the possibility of cold-water immersion (see Figure 5.3); we are assuming that this extends to the scenario of helicopter ditching and egress. For those at daily risk of immersion (e.g., fishers, sports enthusiasts, pilots, aircrew, and passengers in overwater operations), it may be that this preparation should include

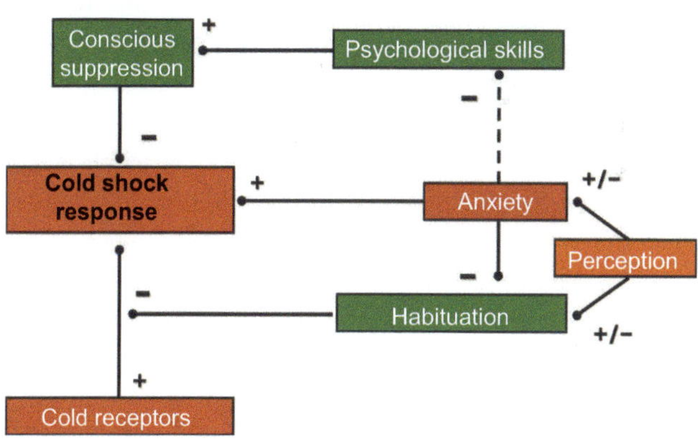

Figure 5.3 Model summarizing the influences of psychological intervention, habituation, anxiety, and perception on the magnitude of the cold shock response. A minus sign indicates that one variable reduces the influence of another, a plus sign indicates that one variable increases the influence of another. Green indicates a positive influence upon the response (i.e., the magnitude of the response is reduced or control over the response is improved), red indicates a negative influence upon the response, and amber indicates a two-way influence whereby the response can be either increased or decreased by a given variable; a dashed line indicates that the relationship shown is untested experimentally.

consideration of the psychological component of the responses to cold water and the benefits that habituation can have in this scenario. Covering these aspects of preparation should not be at the exclusion of specific survival training through simulation such as helicopter underwater escape training or learning to use life-saving appliances such as emergency underwater breathing systems that are designed to support the victim and enable survival; rather a package of coping tools should be developed to reduce the appraisal of threat and improve the coping resource that is available. The resultant effect of considering all of these aspects may be a lowered physiological response in the emergency scenario and an improved chance of survival where circumstances allow. It is also evident that preparatory procedures, such as habituation, can be reversed by coexistence of heightened anxiety on and during cold-water immersion. It is a reasonable assumption that, in a life-threatening situation, anxiety levels will be raised and this may magnify the cold shock response that is seen or interfere with the habituation that has been previously induced. Underpinning many of the studies of the responses to cold water and helicopter underwater escape training is a large inter-individual variation in the responses that are seen. Therefore, the responses to interventions such as PST and habituation cannot be relied upon exclusively. It would certainly be valuable to identify those who respond best to these types of intervention and, in contrast, those who do not; the latter would require greater support in the real-life scenario. The studies that specifically investigate the psychological responses to cold-water immersion and the effects of targeted interventions such as PST are relatively recent, with promising results. More research is required into these effects to maximize the chances of survival in the real-life emergency scenario.

REFERENCES

Bandura, A. (1977). Self-efficacy: toward a unifying theory of behavioral change. *Psychological Review, 87*, 191–215.

Banken, J. A., & Mahone, C. H. (1991). Brief cognitive behavior therapy in an undergraduate pilot student: a case report. *Aviation, Space and Environmental Medicine, 62*(11), 1078–1080.

Barwood. M. J. (2005). *Psychophysiology of survival: The impact of psychological strategies on the physiological responses to thermal environments* (Doctoral thesis). Portsmouth: University of Portsmouth.

Barwood, M. J., Bates, V., Long, G. M., & Tipton, M. J. (2011). "Float First": trapped air between clothing layers significantly improves buoyancy on water immersion in adults, adolescents and children. *International Journal of Aquatic Research and Education, 5*, 147–163.

Barwood, M. J., Corbett, J., Green, R., Smith, T., Tomlin, P., Weir-Blankenstein, L., et al. (2012). Acute anxiety increases the magnitude of the cold shock response before and after habituation. *European Journal of Applied Physiology, 113*, 681–689.

Barwood, M. J., Corbett, J., & Wagstaff, C. R. D. (2014). Habituation of the cold shock response may include a significant perceptual component. *Aviation, Space and Environmental Medicine, 85*, 167–171.

Barwood, M. J., Dalzell, J., Datta, A., Thelwell, R., & Tipton, M. J. (2006). Breath-hold performance during cold water immersion: effects of psychological skills training. *Aviation, Space and Environmental Medicine, 7*, 1136–1142.

Barwood, M. J., Datta, A., Thelwell, R. C., & Tipton, M. J. (2007). Breath-hold performance during cold water immersion: effects of habituation with psychological training. *Aviation, Space and Environmental Medicine, 78*, 1029–1034.

Barwood, M. J., Thelwell, R. C., & Tipton, M. J. (2008). Psychological skills training improves exercise performance in the heat. *Medicine and Science in Sport and Exercise, 40*(2), 387–396.

Brooks, C. J., Gibbs, P. N., Jenkins, J. L., & McLeod, S. (2007). Desensitizing a pilot with a phobic response to required helicopter underwater escape training. *Aviation, Space and Environmental Medicine, 78*(6), 618–623.

Cheung, S. S., D'Eon, N. J., & Brooks, C. J. (2001). Breath-holding ability of offshore workers inadequate to ensure escape from ditched helicopters. *Aviation, Space and Environmental Medicine, 72*, 912–918.

Conroy, D. E., & Metzler, J. N. (2004). Patterns of self-talk associated with different forms of competitive anxiety. *Journal of Sport and Exercise Psychology, 26*, 69–89.

Croft, J. L., Button, C., Hodge, K., Lucas, S. J. E., Barwood, M. J., & Cotter, J. D. (2013). Responses to sudden cold water immersion in inexperienced swimmers following training. *Aviation, Space and Environmental Medicine, 84*, 850–855.

Datta, A., & Tipton, M. J. (2006). Respiratory responses to cold water immersion: neural pathways, interactions, and clinical consequences awake and asleep. *Journal of Physiology, 100*, 2057–2064.

Glaser, E. M., Hall, M. S., & Whittow, G. C. (1959). Habituation to heating and cooling of the same hand. *Journal of Physiology, 146*, 152–164.

Godfrey, S., & Campbell, E. J. M. (1968). Mechanical and chemical control of breath-holding. *Quarterly Journal of Experimental Physiology, 54*, 117–128.

Golden, F. S. C., & Tipton, M. J. (1988). Human adaptation to repeated cold immersions. *Journal of Physiology, 396*, 349–363.

Goode, R. C., Duffin, J., Miller, R., Romet, T. T., Chant, W., & Ackles, A. (1975). Sudden cold-water immersion. *Respiratory Physiology, 23*, 301–310.

Hackett, G., & Horan, J. J. (1980). Stress inoculation for pain: what's really going on? *Journal of Counseling Psychology, 27*, 107–116.

Hackfort, D., & Schwenkmezger, P. (1993). Anxiety. In R. N. Singer, M. Murphey, & L. K. Tennant (Eds.), *Handbook of research on sports psychology* (pp. 328–364). New York: Macmillan.

Keatinge, W. R., & Evans, M. (1961). The respiratory and cardiovascular responses to cold and warm water. *Quarterly Journal of Experimental Physiology and Cognate Medical Sciences, 46*, 83–94.

Lazarus, R. S., & Folkman, S. (1984). *Stress, appraisal, and coping*. New York: Springer Publishing Company.

Leach, J. (1994). *Survival psychology*. London: Macmillan Press Ltd.

Leach, J. (2005). Cognitive paralysis. *Aviation, Space and Environmental Medicine, 76*(2), 134–136.

Levendusky, P., & Pankratz, L. (1975). Self-control techniques as an alternative to pain medication. *Journal of Abnormal Psychology, 84*, 165–169.

Meichenbaum, D. (1972). Cognitive modification of test anxious college students. *Journal of Consulting and Clinical Psychology, 39*, 370–380.

Meichenbaum, D. (1996). Stress inoculation training for coping with stressors. *Clinical Psychology, 49*, 4–7.

Modell, J. H., Gaub, M., Moya, F., Vestal, B., & Swarz, H. (1966). Physiologic effects of near drowning with chlorinated fresh water, distilled water and isotonic saline. *Anaesthesiology, 27*, 33–41.

Ross, M. J., & Berger, R. S. (1996). Effects of stress inoculation training on athletes' postsurgical pain and rehabilitation after orthopedic injury. *Journal of Consulting and Clinical Psychology, 64*(2), 406–410.

Sarason, I., Johnson, J., Berberich, J., & Siegel, J. (1979). Helping police officers to cope with stress: a cognitive behavioural approach. *American Journal of Community Psychology, 7*, 593–603.

Thelwell, R. C., & Maynard, I. W. (2003). Developing competitive endurance performance using mental skills training. *The Sport Psychologist, 4*, 377–378.

Tipton, M. J. (1989). The initial responses to cold-water immersion in man. *Clinical Science, 77*, 581–588.

Tipton, M. J. (1992). The relationship between maximum breath hold time in air and the hyperventilatory responses in cold water. *European Journal of Applied Physiology, 64*, 426–429.

Tipton, M. J. (2003). Cold water: sudden death and prolonged survival. *The Lancet, 362*, S12–S13.

Tipton, M. J., Balmi, P. J., Bramham, E., Maddern, T. A., & Elliot, D. H. (1995). A simple emergency underwater breathing aid for helicopter escape. *Aviation, Space and Environmental Medicine, 66*, 206–211.

Tipton, M. J., Eglin, C. M., & Golden, F. S. C. (1998). Habituation of the initial responses to cold water immersion: a central or peripheral mechanism? *Journal of Physiology, 512*, 621–628.

Tipton, M. J., Franks, C. M., Sage, B. A., & Redman, P. J. (1997). An examination of two underwater breathing aids for use during helicopter underwater escape. *Aviation, Space and Environmental Medicine, 68*, 907–914.

Tipton, M. J., Gibbs, P., Brooks, C., Roiz de Sa, D., & Reilly, T. J. (2010). ECG during helicopter underwater escape. *Aviation, Space and Environmental Medicine, 81*, 399–404.

Tipton, M. J., Mekjavic, I. B., & Eglin, C. M. (2000). Permanence of the habituation of the initial responses to cold-water immersion in humans. *European Journal of Applied Physiology, 83*, 17–21.

Tipton, M. J., & Vincent, M. J. (1989). Protection provided against the initial response to cold immersion provided by a partial coverage wet suit. *Aviation, Space and Environmental Medicine, 60*, 769–773.

Turk. D. (1975). *Cognitive control of pain: A skills training approach for the treatment of pain* (Unpublished Masters thesis). Ontario, Canada: University of Waterloo.

Zald, D. H. (2003). The human amygdala and the emotional evaluation of sensory stimuli. *Brain Research. Brain Research Reviews, 41*, 88–123.

CHAPTER 6

Physiological and Cognitive Changes during Helicopter Underwater Egress Training

Sarita J. Robinson
School of Psychology, University of Central Lancashire, Preston, Lancashire, England

6.1 INTRODUCTION

Helicopter ditching over water is rare, but the survival rate for those who do crash over water is low (Brooks, MacDonald, Baker, Shanahan, & Haaland, 2014; Cheung, D'eon, & Brooks, 2001; Taber, 2014; Taber & McGarr, 2013). Upon crashing, the majority of helicopters will sink rapidly, filling with water, debris, and fuel (Ryack, Luria, & Smith, 1986). Further, many helicopters become unstable upon contact with the water and will invert, leaving survivors strapped in their seats, upside down, and underwater (Taber & McCabe, 2009). As a result, most people involved in a helicopter crash over water do not die of their crash injuries but drown, as they cannot escape from the helicopter structure (Brooks, MacDonald, Donati, & Taber, 2008). The reasons people cannot escape from the helicopter once it has become submerged are complex. Crash events are unfamiliar, time pressured, and can result in rapid psychological and physiological changes in crash victims. All of these factors can cause marked cognitive failures, which in turn can reduce helicopter crash survival.

Training simulators, which can familiarize people with emergency protocols in a safe and controlled environment, have long been used within the aviation industry (Sharkey & McCauley, 1992). Anecdotal reports and the limited research available from helicopter crash survivors suggests that the helicopter underwater egress training (HUET) simulator can play a key role in improving survival (Cunningham, 1978; Ryack et al., 1986). For example, Hytten (1989) reported that four of the five survivors of a helicopter crash into the sea had received simulated helicopter accident training, whereas the one casualty of the crash had not.

The exact way in which simulator training enhances survival is not fully understood. Training may improve survival by allowing people to practice

Handbook of Offshore Helicopter Transport Safety
ISBN 978-1-78242-187-0
http://dx.doi.org/10.1016/B978-1-78242-187-0.00006-7

the use of emergency equipment, such as the emergency rebreather device, which has been found to increase helicopter crash survival rates (Brooks, Muir, & Gibbs, 2001; Taber & McCabe, 2009). The provision of survival equipment is of little use if people have not been previously trained to use the equipment provided (Taber, 2013). Further, training can help to reduce the common problems seen in helicopter underwater egress, such as the inability to release seat belts or locate and open exits (Ryack et al., 1986). Finally, training can help in familiarizing people with the crash environment, enhancing the individuals' confidence that they are able to handle an emergency situation (Hytten, Jensen, & Vaernes, 1989; Ryack et al., 1986). However, a more detailed investigation into how people respond to HUET on a physiological and cognitive level could lead to a better understanding of: 1) possible cognitive failures that may occur in real-life crash situations, 2) factors that could have an impact on survival, and 3) improvements that could enhance training. Therefore this chapter explores physiological and cognitive changes that occur in response to HUET and considers how our understanding of these changes can inform the HUET program.

6.2 TRAINING RELUCTANCE

HUET is a standard requirement for working offshore. However, some offshore workers may not fully engage in the training. This can be for several reasons.

- One reason that offshore workers may not engage in HUET is that they either do not understand the risk posed by flying over water or believe that a helicopter crash over water would be nonsurvivable. The lack of a realistic understanding of the risks posed by a potential disaster situation has been shown to lead to a failure to engage with emergency procedures. For example, residents living near Mount St. Helens were found not to have evacuated the local area as they did not understand the risks associated with a volcanic eruption (Greene, Perry, & Lindell, 1981). Although the survival rates for helicopter crashes over water are not good, with the most current survival rates being suggested to be around 68% (Taber, 2014; Taber & McCabe, 2006), training has been shown to enhance survival (Ryack et al., 1986). Therefore it is important that HUET outlines the potential risks of helicopter flights over water and the benefits of training on enhancing survival rates to facilitate engagement.

- Another reason people may not engage fully with the HUET is that they are using the coping strategy of denial to deal with their anxiety regarding helicopter flights. Refusing to believe that a helicopter crash is possible can be psychologically protective as it reduces the anxiety associated with flying (Tobin, Holroyd, Reynolds, & Wigal, 1989). If a person does not accept that helicopters can crash then there is no need to worry about the possibility. However, denial can also lead to maladaptive coping, as people who deny risks may then not fully engage in useful proactive activities (Grothmann & Patt, 2005) such as HUET. Therefore at the start of HUET courses it is important to assess the participants' thoughts regarding the likelihood of helicopter ditching and ensure that participants understand the benefits of safety training.
- A final reason people may fail to engage with HUET is the increased anxiety levels that the training may induce. People who have a preexisting phobia of water or fear of enclosed spaces may find HUET especially challenging. If people perceive that they are at risk of physical harm from undertaking an activity it is likely that they will not be able to fully engage with that activity (Lazarus, 1966). Therefore, training courses should not underestimate the negative effects of anxiety on HUET. Interventions to reduce the anxiety that HUET can evoke are possible. For example, systematic desensitization (a common treatment for phobias) has been found to reduce phobic anxiety associated with HUET (Brooks, Gibbs, Jenkins, & McLeod, 2007). If the anxiety is less marked, interventions such as prepool training using virtual reality technologies (Reznek, Harter, & Krummel, 2002) or techniques such as goal-setting and positive self-talk may be helpful (Barwood, Dalzell, Datta, Thelwell, & Tipton, 2006).

6.3 PHYSIOLOGICAL CHANGES

During our everyday life the brain continually scans the environment for signs of danger (Eilam, Izhar, & Mort, 2011). If a part of the environment is defined as being dangerous then that sensory input is passed to a part of the brain called the amygdala, which is responsible for the processing of emotional information. See Figure 6.1 for a labeled diagram of the relevant brain structures.

When a person is faced with HUET (Figure 6.2), his or her appraisal of the situation will determine whether he or she defines the training as a threat (see Chapters 1 and 7 on training for more detail).

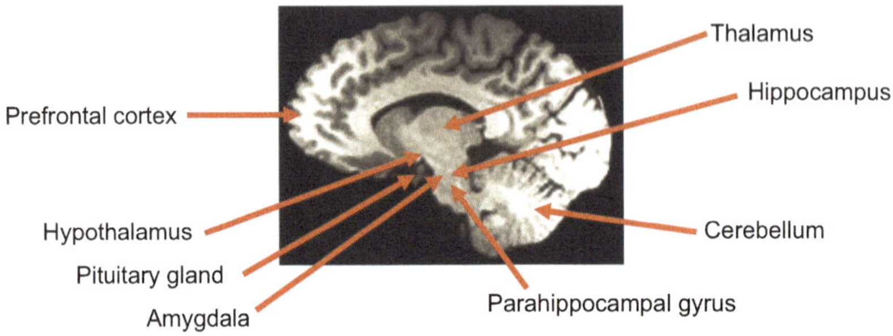

Figure 6.1 Diagram of key brain structures. *(Image provided by Dr. Faye Skelton.)*

Figure 6.2 The helicopter underwater egress training vehicle at Fleetwood Nautical College, England.

Course participants who are confident in their ability to complete HUET are less likely to see the training as a threat and so activation of the amygdala is less likely to occur. Participants may have enhanced confidence in relation to HUET if they have successfully completed the training previously (Taber & McGarr, 2013) or they already have a high degree of confidence in the water (e.g., divers). However, for others the HUET may be processed as being a significant threat and so the threat information will be passed to the amygdala. Once the amygdala has confirmed the presence of a threat, the hypothalamus will trigger the two main physiological systems that are designed to enhance self-preservation by helping the body to cope with the danger (Miller & O'Callaghan, 2002; Selye, 1974). The first system to be activated is the sympathomedullary pathway (SAM), which is responsible for the short-term fight or flight response (see Figure 6.3).

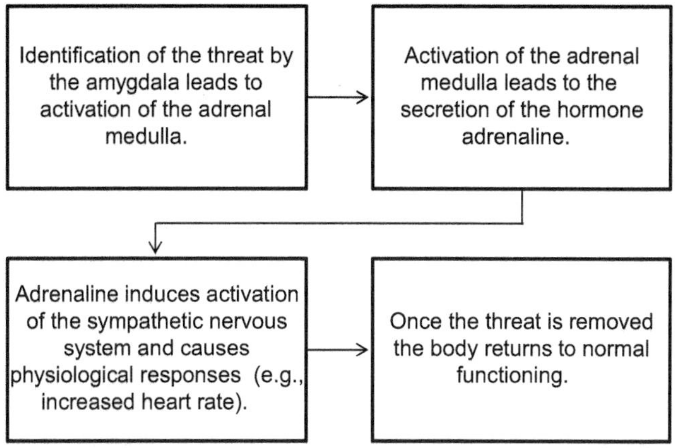

Figure 6.3 Graphical illustration of the sympathomedullary pathway.

The actions of the SAM pathway lead to a number of physiological changes that help the body to deal with an environmental threat. For example, to deal with a physical threat the body may need more oxygen to fuel a "fight or flight" response. Activation of the SAM pathway triggers the airways in the lungs to expand so that the body can take in an increased amount of oxygen. The increase in levels of adrenaline that results from SAM activation triggers the release of glucose from the liver and temporary fat stores so that the body has more energy to deal with the perceived danger. In order that the increased supply of oxygen and glucose can reach the brain and muscles faster, activation of the SAM pathway also leads to the heart beating faster to allow the blood to be pumped more quickly around the body.

At the same time as the SAM pathway is activated the hypothalamus also activates the slower acting hypothalamic–pituitary–adrenal (HPA) axis. The HPA axis is designed to maintain a more sustained response to threat (see Figure 6.4).

The HPA axis helps to maintain a sustained stress response if the brain continues to signal that a threat is present. If the threat is deemed to still be present the hypothalamus will secrete corticotropin-releasing hormone (CRH). CRH travels to the pituitary gland and triggers the release of adrenocorticotropic hormone (ACTH). ACTH then causes the adrenal glands (located above the kidneys) to increase the levels of circulating cortisol in the blood. Cortisol helps the body to generate the resources needed for a sustained response to threat, such as maintaining high levels of

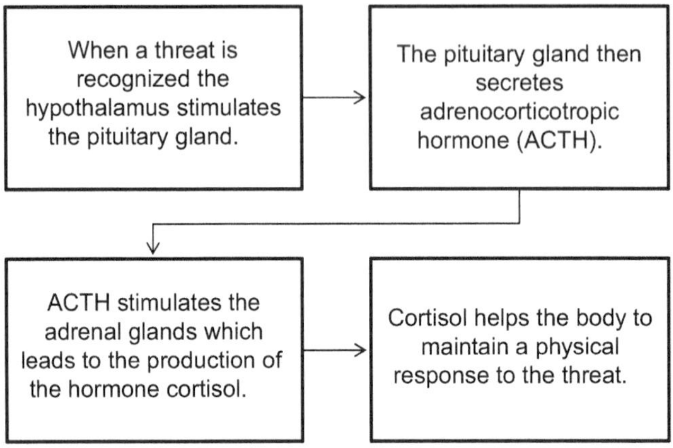

Figure 6.4 Graphical illustration of the hypothalamic–pituitary–adrenal axis.

blood glucose. If cortisol levels remain high, then an appropriate stress response is maintained. If the levels of circulating cortisol reduce, then the threat is deemed to have been removed and the body returns to normal functioning.

Activation of the SAM pathway and the HPA axis can help the body to elicit a physical response to the perceived threat from the HUET. However, the associated neurochemical changes can have a negative impact on cognitive processes, such as memory and attention. For example, in addition to the fight or flight response, Leach (2004, 2005) has suggested that people can suffer from cognitive paralysis (also known as freezing) during acutely stressful situations (see Chapter 2 for further details).

Both the SAM pathway and HPA axis activation have been shown to lead to fluctuations in certain neurochemicals and can lead to changes in regional cerebral blood flow. For example, exposure to acute stressors can cause changes in the levels of catecholamine, neurotransmitters, and adrenaline within the brain (Thomas, Lockwood, Singh, & Deuster, 1999). In addition, acute stress can significantly increase the levels of circulating cortisol within the body (Dickerson & Kemeny, 2004; Robinson, Sünram-Lea, Leach, & Owen-Lynch, 2008; Sünram-Lea, Owen-Lynch, Robinson, Jones, & Hu, 2012). Cortisol can then cross the blood/brain barrier and cause impairments in cognition (Buss, Wolf, Witt, & Hellhammer, 2004; Kirschbaum, Wolf, May, Wippich, & Hellhammer, 1996; Lupien, Gillim & Hauger, 1999). Therefore, people who see the HUET as a threat may well show cognitive impairments as a result of the body's physiological response to the perceived stressor. Cognitive impairments during HUET may not only influence the usefulness of the training but may also be linked to an increased risk of injury. The next section outlines some of the potential cognitive impairments that may result from a physiological stress response not only during HUET but also during real-life helicopter emergencies.

6.4 HUET AND COGNITION

Successful egress after a helicopter crash into water, both in real life and during simulator training, demands that complex goal-directed behaviors are carried out. Cognition can be impaired as a result of the physiological changes in response to threat, which in turn can induce neurochemical changes. However, situational factors, such as increases in novel information and the time limitations available to generate useful behaviors during an emergency, can also affect people's ability to undertake useful survival behaviors (Leach, 2004, 2005; Robinson et al., 2008). In addition, other factors that can also be present during an emergency, such as physical fatigue, can also induce a broad spectrum of cognitive decrements (Lorist, Boksem, & Ridderinkhof, 2005).

6.4.1 Prefrontal Cortex

Physiological responses to stressors, such as HUET, especially when coupled with adverse environmental pressures, such as increases in novel stimuli (Ashcraft & Kirk, 2001; Miyake & Shah, 1999), have been found to have an impact on the functioning of an area of the brain known as the prefrontal cortex (PFC; Arnsten, 2009). Specifically, increases in the circulating levels of catecholamine (Brozoski, Brown, Rosvold, & Goldman, 1979), dopamine and noradrenalin (Arnsten & Goldman-Rakic, 1998; Banderet & Lieberman, 1989; Dollins, Krock, Storm, Wurtman, & Lieberman, 1995), and cortisol (Qin, Hermans, van Marle, Luo, & Fernández, 2009) have been shown to influence the functioning of the PFC.

The PFC is the anterior part of the frontal lobes, which are located at the front of the brain. The PFC plays a number of critical roles and mainly controls higher level thinking such as social regulation and decision-making (Arnsten, 2009). One key function of the PFC is to support working memory (WM). WM monitors the external world continuously and is responsible for manipulating, suppressing, and integrating information to allow the brain to direct appropriate behavior (Funahashi, 2001).

People exposed to acutely stressful stimuli exhibit PFC dysfunction, which in turn has been found to negatively affect WM performance

(Leach & Griffith, 2008; Porter & Leach, 2010). Increases in anxiety levels may cause WM impairments as WM undertakes task-irrelevant processing, such as dealing with worries and self-concern; this means that there is less WM available for more appropriate tasks (Darke, 1988; Eysenck & Calvo, 1992). WM impairments during HUET are likely to have an impact on training effectiveness and students may need specific modifications to the training to aid completion (see Chapter 7 for more detail on training modifications to help individuals who have difficulty in completing HUET).

Impairments in WM due to lesions in the prefrontal lobes have been shown to lead to difficulties in organizing information (Shimamura, Janowsky, & Squire, 1990) and monitoring the temporal sequences of events (Milner, 1982). Failure in the WM during HUET may therefore lead to problems in ordering action sequences. For example, course participants may not carry out the actions required to use the rebreather in the correct order. Of course participants may try to egress through the helicopter window before they have undone their seat belt. Therefore, course participants may need to be repeatedly reminded of the order in which to carry out tasks.

During HUET, problems with decision-making and behavior generation may also be observed as a result of impairments in WM. Processing large amounts of novel information in a short period is difficult if WM resources are not available (Janis & Mann, 1977; Pires, 2005). If WM is limited, then the individual becomes more selective about which information to process (Lavie, 1995). This can lead to the neglect or forgetting of important information (Payne, Bettman, & Johnson, 1993) and the use of strategies that are less demanding, less time-consuming, but also less accurate (Ozel, 2001). This in turn can lead to a higher probability of error and increased instances of erroneous action (Ariely & Zakay, 2001; Entin & Serfaty, 1990). Therefore, during HUET it can be difficult for course participants to decide on a course of action and then generate and execute the new behavior (Cowan, 1999). As a result of impairments in WM, some course participants may not be able to generate any behavioral response and so may freeze (Leach, 2004, 2005) or generate an inappropriate response (Leach, 1994). To support participants in their decision-making and behavior construction during HUET, course instructors should offer clear and concise instructions. Further, course instructors should allow time for new complex procedures to be practiced several times before being completed in the HUET. For example, additional training in the use of the emergency rebreather device prior to HUET can increase the use of the equipment during egress (Taber, Kozey, & McCabe, 2012).

Another key function of WM that may be impaired as a result of HUET is inhibition control. Inhibition control refers to WM ability to screen out nonrelevant information so that the contents of WM are restricted to information that is relevant to the task being carried out (Rosen & Engle, 1998). Exposure to stressors can lead to failures in inhibition control (Hartley & Adams, 1974) and impairments of inhibitory control in people undergoing HUET have been observed. During HUET at the Nautical College in Fleetwood, England, failure to release the seat harness seemed to occur because course participants tried to release their harness as if it were a car seat belt. Some course participants appeared unable to inhibit the automatic movement associated with releasing a car seat belt and so repeatedly pressed the side of their seat instead of pressing the release mechanism in the middle of their abdomen. Failures in inhibition control in people undertaking HUET have also been observed by Robinson, Leach, and Walsh (2009). This study asked participants to complete the Stroop task immediately after egress. In the Stroop task participants are presented with a number of color names (such as red, blue, or green) written in different-colored inks. The participants are then asked to ignore the color name and instead say the color of the ink. For example, if the word "red" is written in green ink, the participant would respond "green." Therefore the participant is being asked to inhibit his or her automatic tendency to read the word and instead say the ink color. Robinson et al. (2009) found that immediately after egress during HUET participants showed a marked impairment in their Stroop performance, indicating impairment in inhibition control.

Further, the lack of standardization in helicopter door and window design (Brooks & Bohemier, 1997) is likely to lead to inhibition control problems. Course participants are likely to become familiar with one design of helicopter door/window and so may not be able to inhibit the automatic actions associated with the door/window that they most often use (Taber, 2014). The inability to inhibit the automatic door-opening action may lead to a delay or failure in egress. Until a universal jettison mechanism for doors and windows, as proposed by Brooks, Miller, Morton, and Baranski (1999), is adopted by all helicopter manufacturers, problems with inhibition control may well lead to failures in egress. Until a universal jettison mechanism is adopted, HUET participants should be given the opportunity to practice with the type of door/window mechanism that will be most common to the type of helicopter that they will be flying in (Taber, 2014; Wells, 2010).

Overall, it should be noted that one of the most effective ways of reducing the negative impact of physiological responses to threat on cognition is to reduce demand on areas of cognition that are particularly susceptible. As WM has been shown to become impaired during threat (Robinson et al., 2008, 2009), then any training that can reduce demand on the WM during a helicopter emergency may aid survival. Therefore the HUET program may be able to help improve survival in two key ways. First, training can reduce the burden on WM by encoding procedural memories, which can be triggered during an emergency and recalled without the need for WM involvement. Procedural motor memories are encoded into a long-term memory store located in the cerebellum (Nezafat, Shadmehr, & Holcomb, 2001; see Figure 1) and triggered automatically during an emergency. For example, learning how to automatically use an emergency rebreather device means that the brain can carry out a useful action during a helicopter crash with minimal input from the prefrontal cortex (where WM is located). Therefore, HUET should be repeated until participants have had sufficient practice to have laid down the motor memories needed to complete a successful underwater egress without conscious thought. Further, training should be as close to a realistic scenario as possible. Taber and McGarr (2013) note that currently training for emergency breathing system (EBS) use involves practicing deployment only on the surface prior to aircraft inversion. No training is given on how to deploy the EBS after inversion. This means that during a real helicopter ditching people will have to draw on their limited WM to create the behavior needed to deploy the EBS underwater. Limited WM during the ditching may result in a failure to generate the behavior needed to deploy the EBS or an incorrect but similar behavior being carried out (Leach, 2004, 2005). Therefore it is essential that training protocols match the most likely emergency scenarios and that HUET allows overlearning of the protocols needed for successful underwater egress. In addition, it has been suggested that retention of learning from the HUET should be tested after a period of consolidation, and that HUET should be carried out frequently to ensure that emergency protocols are retained (Taber & McGarr, 2013).

The second key way in which training reduces WM impairment is by helping people to feel more confident in their ability to survive a crash situation (Taber & McGarr, 2013). If the training increases people's confidence levels then physiological responses and associated anxiety and cognitive impairments are likely to be less marked during exposure to a real-life stressful event (Deladisma et al., 2009). Therefore, it is critically important that all people undertaking the HUET program leave with a confident but realistic view of their ability to carry out the survival procedures effectively.

6.4.2 Hippocampus

Another area of the brain that can be adversely affected by the physiological changes that occur when the body is placed under threat is the hippocampal structures. The hippocampus has been shown to be important for the encoding and storage of declarative memories (Scoville & Milner, 1957), with the parahippocampal gyrus being especially important for long-term spatial memory (Maguire, Frackowiak, & Frith, 1997). Declarative memories are considered to be knowledge-based information, for example, our ability to recall the names of the capital cities of different countries. Although HUET aims to teach procedural memories (unconsciously recalled motor skills) such as how to automatically deploy an EBS, people undergoing HUET need to be able to access declarative memories to hold and recall essential instructions during skill acquisition.

During threat, increases in the hormone cortisol may be one of the reasons for failures in declarative memory (Buss et al., 2004; De Quervain, Roozendall, Nitsch, McGaugh, & Hock, 2000; Kirschbaum et al., 1996; Lupien, Gillim, & Hauger, 1999; Lupien & McEwen, 1997; Newcomer et al., 1999; Tops, van der Pompe, & Wijers, 2004), with larger increases in cortisol being associated with greater declarative memory impairments (Wolf, Schommer, Hellhammer, Reischies, & Kirschbaum, 2002). As higher levels of cortisol are normally associated with a more negative response to the HUET, course participants who are more anxious about undertaking the training are likely to show declarative memory failures.

Impairments in encoding declarative memories have been observed in people undergoing the HUET (Robinson et al., 2008) and so instructors can expect HUET course participants to need more frequent reminders regarding how to complete the emergency procedures they are learning. Further, training may take more time and be more effortful for people who are experiencing elevated levels of anxiety during the training. Therefore, learning may well be more effective during HUET if anxiety levels are kept to a minimum. Course instructors may also find it useful to include a degree of overlearning (repeating the training even though the emergency response appears to have been successfully encoded) to ensure that the memories are available for later recall. Further strategies, such as the use of mnemonics, can support memory processes and so enhance learning. For example, the Fleetwood Nautical Campus instructors successfully use the mnemonic S.H.I.T. (sharp objects, hood, inflatable, touchdown) to improve recall of the precrash safety drill (personal communication, Joe Bottomley, Head of Offshore Operations, Fleetwood Nautical Campus, 2013).

6.4.3 Individual Differences in Response to Threat

Although physiological responses to threat are commonly observed, the exact response observed in an individual is subject to individual differences. For example, research into the neurotransmitter neuropeptide Y (NPY) in soldiers undergoing military survival training indicated that those soldiers who had higher levels of NPY displayed better behavioral responses and showed less distress following exposure to an uncontrollable acute stressor (Morgan et al., 2000). Further, the androgen steroid hormone dehydroepiandrosterone (DHEA), which is triggered by ACTH during stress exposure, has also been found to be an important factor in how people tolerate exposure to stress. Research indicates that the ratio between DHEA and glucocorticoids, such as cortisol, appears to be an important factor in how people tolerate exposure to stress and the extent to which cognitive impairments are observed. A higher ratio between DHEA and cortisol seems to provide a buffer against the damaging effects of stress and reduce the likelihood of people in survival situations experiencing cognitive deficits (Morgan et al., 2004). This raises the interesting question as to whether some people may have a genetic predisposition to be able to better cope with adverse responses during threat. Alternatively, psychosocial factors, such as social support or mood state, may mediate the stress response and so influence an individual's physiological response to stress.

6.5 MEDIATORS OF THE PHYSIOLOGICAL RESPONSE TO HUET

Higher levels of anxiety during HUET appear to elicit a greater physiological response, which in turn appears to lead to more neurochemical changes and so result in more marked cognitive impairments (Robinson et al., 2008). The appraisal of the HUET by the brain is therefore the key to determining the physiological response made to the situation. If anxiety levels can be kept to a minimum, then it might be possible to stop some of the cognitive impairments from occurring. A number of factors have been suggested that could act as useful interventions in reducing the negative physiological and psychological responses to HUET.

6.5.1 Social Support

Social support has been shown to have a powerful influence on how we respond to threat (Cohen & Wills, 1985). In everyday life, studies have found that people with spouses, friends, and family members who can provide psychological and material resources have better mental and physical health (Barth, Schneider, & von Känel, 2010; Pinquart & Duberstein, 2010). Social support can be divided into two main forms. The first is informal social support, which refers to personal networks of family and friends who can help out in practical and emotional ways. The second is formal social support, which refers to the support that can be gained from paid professionals from social and community structures, such as doctors and teachers.

One theory suggests that social support has a beneficial effect on anxiety as it acts as a "buffer" protecting people from negative events (Cohen & Wills, 1985). The importance of social support in helping people to cope with real-world stressors has often been shown after exposure to trauma. For example, Ren, Skinner, Lee, and Kazis (1999) found that people with better social support networks before and after exposure to a traumatic event, such as a motor vehicle accident or a physical assault, had better outcomes 12 months later. During exposure to a stressful situation, social support appears to help people appraise the stressor as being less threatening and so physiological responses, such as increases in blood pressure, are less marked (Roy, Steptoe, & Kirschbaum, 1998). For example, men undertaking a public speaking task reported less stress and had lower cortisol responses when in the presence of a friend (Heinrichs Baumgartner, Kirschbaum, & Ehlert, 2003; Kirschbaum et al., 1995). Other studies have also found benefits from having a friend present during exposure to a psychological stressor (such as simulated job interview or a discussion on a controversial topic), with the presence of social support resulting in less cardiovascular reactivity (Gerin, Pieper, Levy, & Pickering, 1992; Kamarck, Manuck, & Jennings, 1990). The suggested buffering action of social support may work by helping people to interpret stressful events in a more positive light (Uchino, 2006) or by giving people the confidence to cope with adverse situations (Gottlieb & Bergen, 2010). Therefore course instructors should encourage course participants to undertake training with supportive friends or work colleagues if they appear to be anxious about the HUET. If this is not possible, course instructors may need to act as a formal social support network and offer additional help and encouragement to course participants who appear to be struggling.

Interestingly, the positive effects of social support for an individual during exposure to a stressor are seen not only when a friend or family member is physically present. Research has suggested that social support levels can influence people's stress responses to daily hassles even when they are not in the direct presence of their friends (Gottlieb & Bergen, 2010). This finding was supported by Robinson, Leach, and Bridges (2010), who looked at the benefit of social support on HUET. Course participants undertaking the HUET exhibited low cortisol levels and subjective anxiety prior to the training if they reported high levels of social support in their home lives. Therefore, in addition to the social support available during training, the course participant's everyday level of social support may also be important. If a course participant has recently experienced bereavement or is undergoing a divorce or is in any other situation in which his or her social support network has become impaired, he or she should consider delaying HUET.

6.5.2 Mood State

Mood state has also been found to have an impact on people's performance in high stress situations. For example, low mood (having a number of depressive symptoms but not necessarily at a clinical level for a diagnosis of depression) can lead to poorer cognition generally (Cohen, Weingartner, Smallberg, Pickar, & Murphy, 1982) and can also affect how people approach stressful situations (Ravindran, Matheson, Griffiths, Merali, & Anisman, 2002). For example, in Ravindran et al. (2002), depressive adults reported more day-to-day stressors, had a reduction in the amount of perceived uplifting events, and reported a poorer quality of life.

In contrast, a positive outlook on life has been found to be beneficial to people both in everyday life (Scheier & Carver, 1987) and in high stress situations (Kivimäki et al., 2005). Being an optimist (having a general expectation that life will be positive) has been shown to influence our physiology. Segerstrom, Taylor, Kemeny, and Fahey (1998), for example, found that optimists had enhanced immune function compared to those who were more pessimistic. Segerstrom et al. (1998) speculated that this was because optimists were more positive in the way in which they coped with stressors, as they experienced fewer negative moods and undertook more adaptive behaviors. Therefore, mood state and general outlook on life may well modulate how well people undergoing the HUET interpret and cope with the stress. Those with a low mood and a pessimistic outlook may be more likely to find the HUET challenging. Interventions that improve mood or increase optimism, such as the Seligman (2011) "Learned Optimism" program, may be useful prior to training. On the day of training encouraging course participants to think positively about the future (Fosnaugh, Geers, & Wellman, 2009) may enhance situational optimism about HUET and so increase training completion.

High levels of optimism and unrealistically high levels of confidence are not always beneficial, however. If people become unrealistically optimistic about their ability to deal with an emergency, they may feel that the survival training is not important. For example, Weinstein (1987) found that people who were highly optimistic were less likely to reduce behaviors that pose a health risk, such as smoking. It could well be that people with high levels of optimism may not fully engage with the HUET as they do not believe that they are likely to be exposed to a helicopter crash over water. Therefore, it is critical to strike the right balance between giving people the confidence that they can complete the training and stopping people from becoming overconfident and so feeling that the training is unnecessary.

6.5.3 Nutrition

The nutritional status of people exposed to acute stressors, such as HUET, can affect the extent to which cognitive failures occur. Training that includes high physical demand, such as HUET, can lead to a significant decrease in blood glucose levels (Smith, Petruzzello, Chludzinski, Reed, & Woods, 2001) and increased dehydration (Cian, Koulmann, Barraud Raphel, Jimenez, & Melin, 2000), both of which have been shown to impair cognition (Sünram-Lea et al., 2012). Therefore, adequate hydration and a good level of blood glucose may lead to better cognition (Smith, Riby, Eekelen, & Foster, 2011). Further, cortisol levels can also become heightened as a result of HUET (Robinson et al., 2008). Because cortisol is released to ensure that the body has an adequate supply of glucose during threat, people with a relatively high blood glucose level should not have such a marked increase in circulating cortisol. This being the case, HUET is likely to be more successful if students on the course are encouraged not to restrict their food intake. Further, people who appear especially anxious should be encouraged to take on additional glucose prior to training.

As well as maintaining blood glucose levels, course participants should not reduce their level of common legal stimulants, such as nicotine and caffeine. It has been observed that people who are attempting to give up smoking and so enter nicotine withdrawal can have impaired cognition (Xu et al., 2005). Similar individuals who are experiencing caffeine withdrawal can show decreases in their level of alertness (Schuh & Griffiths, 1997). In addition, caffeine has been found to have a positive effect on cognition (Jarvis, 1993), even in highly stressful environments (Lieberman, Tharion, Shukitt-Hale, Speckman, & Tulley, 2002). As a result of these observations, it may be sensible for people to not attempt to give up smoking or drinking caffeinated beverages shortly before completing HUET. Further, smoking and coffee breaks prior to training might be beneficial for students' cognitive performance during HUET.

6.5.4 Drug Enhancers for Performance

Legal cognitive enhancers, such as the stimulant caffeine, have been used over many centuries to improve cognitive performance. More recently, there has been an increase in prescription drugs being used by nonclinical populations to improve cognitive functioning. Drugs such as Ritalin (methylphenidate) used in the treatment of attention deficit disorders have been used to increase WM performance via their actions on the catecholamine system. Other drugs, such as Aricept (donepezil) used by Alzheimer disease sufferers, can improve memory by raising the levels of acetylcholine in the brain in healthy individuals. Studies that have looked at the benefits of using cognitive-enhancing drugs in stressful situations have found some real benefits. For example, commercial pilots who took the drug donepezil for 30 days were able to deal with flight emergencies in a simulator better than those who had taken a placebo (Yesavage et al., 2002). Although no studies have yet been undertaken to examine the impact of cognitive enhancers on HUET or performance during real emergencies, it is a field of research that holds much promise.

6.6 REALISM IN HUET

Survival training is sometimes criticized for not being as demanding as an actual survival situation. Further, it has been suggested that the training would be much more effective if it took place under more realistic conditions (Civil Aviation Authority, 2014; Taber, 2013, 2014; Wells, 2010). For example, it has been suggested that HUET should take place in cold water rather than the more pleasant swimming pool temperatures of around 30 °C. HUET in cold water would be more realistic and would expose participants to the harsh reality of what it would be like to ditch in an area like the North Sea. However, the adverse reaction—both physiologically and psychologically—to the cold water would induce conditions that would not be conducive to learning (Mäkinen, 2007) and may cause negative physical responses such as an increase in cardiac arrhythmias (Tipton, Gibbs, Brooks, Roiz de Sa, & Reilly, 2010). The point of HUET is not to allow the person to experience the full impact of what a helicopter crash on water would be like, but to enable people to develop the skills needed to evacuate a submerged helicopter if they are unlucky enough to find themselves in that situation in the future. Therefore, making HUET more demanding by having the training take place in cold water or including a more rapid, uncontrolled descent into the pool would not be beneficial to learning. The negative physiological and psychological responses induced by a more hostile training environment would impair learning. Therefore, HUET should be realistic enough to allow participants to practice the use of emergency survival equipment and to build their confidence in dealing with emergency situations. The HUET environment should not, however, be so hostile that it causes increases in anxiety or other negative responses, which stop learning from taking place. See Chapter 7 for further details on how to design an effective HUET environment.

6.7 PHYSIOLOGICAL RESPONSES POSTTRAINING

Exposure to acutely stressful emergency situations can lead to impairments in cognition during the event (Porter & Leach, 2010; Robinson et al., 2008; Sünram-Lea et al., 2012). However, the negative effects of acute stress on human physiology and cognition are transient, with the majority of people returning to normal functioning with an hour or so after the environmental stressor subsides (Dickerson & Kemeny, 2004;

Baum, Fleming, & Davidson, 1983; Robinson, Leach, Owen-Lynch, & Sünram-Lea, 2013). Yet, after the immediate threat from a stressor, whether a real-life threat or simulated, some people who appeared cognitively intact during the actual event may show signs of psychological and physiological collapse. For example, medical staff who have observed disaster victims have noted that survivors often become suggestible and passive after rescue (Baker & Chapman, 1962). This poststressor cognitive dip has also been observed after emergency safety training. Participants undertaking a simulated firefighter search and rescue exercise showed no cognitive impairments when tested immediately after the training. However, when course participants were tested 20 min later cognitive impairments were evident (Robinson et al., 2013).

Passive behaviors and the impairments in cognition exhibited after exposure to a major stressor, such as HUET, may be the result of physical and psychological exhaustion, as energy levels have been depleted (Dietrich & Sparling, 2004; Lorist et al., 2005). A reduction in cognitive resources after the initial danger has passed may leave people in real-life emergencies vulnerable to secondary dangers (Robinson et al., 2013). During training, students, especially those who may be undertaking emergency rescue work in the future, should be made aware of the potential for a drop in cognition to occur after the initial emergency has abated. HUET instructors should make course participants aware that they may suffer a dip in cognition post-HUET. Therefore, instructors should discourage participants from undertaking cognitively demanding tasks, such as driving home, immediately following HUET.

6.8 RECOMMENDATIONS

- At the start of training instructors should ensure that all course participants are fully aware of the potential risks posed by flying over water in a helicopter. Instructors should also clearly outline the benefits of HUET for enhancing crash survival rates.
- Anxiety levels of course participants should be assessed prior to training. People who appear overly anxious should be offered an appropriate intervention to help reduce their anxiety levels. Lower levels of anxiety during HUET facilitate better retention of the training program.
- Physiological reactions and the related psychological changes as a result of HUET can lead to marked cognitive impairments, especially those

governed by the PFC and hippocampus within the brain. As a result, impairments in WM and declarative memory may be observed in course participants during training. Therefore course instructors should:

  - Repeat instructions regularly as course participants' ability to retain and process information may be compromised;
  - Break down the order of instructions for activities such as the pre-crash safety drill into easy-to-remember stages;
  - Students may suffer from inhibition control issues, which could result in behaviors such as trying to unclip their harness at the side (like a car seat belt) rather than in the middle of their chest. Course instructors should encourage the students to practice actions that are commonly completed incorrectly several times prior to carrying them out underwater.

- A lack of social support on the course or at home (e.g., owing to bereavement or a divorce) might lead to a higher level of anxiety during training. If course participants are especially anxious about completing HUET, then the training should be avoided if possible until home-life situations are stable and/or a friend can also attend the training. If training needs to be undertaken, then extra support and encouragement should be given by course instructors.
- Participants who suffer from a low mood state or have a pessimistic outlook on life may find the HUET more challenging. Interventions, such as thinking positively about the outcome of the training, may improve situational optimism and so increase the likelihood of training being completed successfully.
- Opportunities should be available for course participants to have a drink and something to eat prior to training so that blood glucose and hydration levels are optimal.
- Course participants should have an opportunity for a cigarette break and to have a cup of coffee prior to training. Low levels of nicotine can impair cognition in smokers and caffeine has been showed to work as a cognitive enhancer.
- HUET should always be realistic enough that participants can learn in a "hands-on" manner the emergency procedures needed to egress from a submerged helicopter. However, the training should not be so realistic that high anxiety or negative physiological reactions (e.g., to cold water) stop learning from taking place.

REFERENCES

Ariely, D., & Zakay, D. (2001). A timely account of the role of duration in decision making. *Acta Psychologica, 108*(2), 187–207.

Arnsten, A. F. (2009). Stress signalling pathways that impair prefrontal cortex structure and function. *Nature Reviews Neuroscience, 10*(6), 410–422.

Arnsten, A. F., & Goldman-Rakic, P. S. (1998). Noise stress impairs prefrontal cortical cognitive function in monkeys: evidence for a hyperdopaminergic mechanism. *Archives of General Psychiatry, 55*(4), 362.

Ashcraft, M. H., & Kirk, E. P. (2001). The relationships among working memory, math anxiety, and performance. *Journal of Experimental Psychology: General, 130*(2), 224.

Baker, G. W., & Chapman, D. W. (Eds.). (1962). *Man and society in disaster.* New York: Basic Books.

Banderet, L. E., & Lieberman, H. R. (1989). Treatment with tyrosine, a neurotransmitter precursor, reduces environmental stress in humans. *Brain Research Bulletin, 22*(4), 759–762.

Barth, J., Schneider, S., & von Känel, R. (2010). Lack of social support in the etiology and the prognosis of coronary heart disease: a systematic review and meta-analysis. *Psychosomatic Medicine, 72*(3), 229–238.

Barwood, M. J., Dalzell, J., Datta, A. K., Thelwell, R. C., & Tipton, M. J. (2006). Breath-hold performance during cold water immersion: effects of psychological skills training. *Aviation, Space, and Environmental Medicine,* 77(11), 1136–1142.

Baum, A., Fleming, R., & Davidson, L. M. (1983). Natural disaster and technological catastrophe. *Environment and Behavior, 15*(3), 333–354.

Brooks, C. J., & Bohemier, A. P. (1997). Helicopter door and window jettison mechanisms for underwater escape: ergonomic confusion! *Aviation Space Environmental Medicine, 68*(9), 844–857.

Brooks, C. J., Gibbs, P. N., Jenkins, J. L., & McLeod, S. (2007). Desensitizing a pilot with a phobic response to required helicopter underwater escape training. *Aviation Space Environmental Medicine, 78,* 618–623.

Brooks, C. J., MacDonald, C. V., Baker, A. P., Shanahan, D. F., & Haaland, W. L. (2014). Helicopter crashes into water—Warning time, final position, and other factors affecting survival. *Aviation, Space, and Environmental Medicine, 85*(4), 440–444.

Brooks, C. J., MacDonald, C. V., Donati, L., & Taber, M. J. (2008). Civilian helicopter accidents into water: analysis of 46 cases, 1979–2006. *Aviation, Space, and Environmental Medicine, 79*(10), 935–940.

Brooks, C. J., Miller, L., Morton, S., & Baranski, J. (1999). Evaluation of a new universal jettison mechanism for helicopter underwater escape. *Aviation, Space, and Environmental Medicine, 70*(8), 752–758.

Brooks, C. J., Muir, H. C., & Gibbs, P. N. G. (2001). The basis for the development of a fuselage evacuation time for a ditched helicopter. *Aviation, Space, and Environmental Medicine, 72*(6), 553–561.

Brozoski, T. J., Brown, R. M., Rosvold, H. E., & Goldman, P. S. (1979). Cognitive deficit caused by regional depletion of dopamine in prefrontal cortex of rhesus monkey. *Science, 205*(4409), 929–993.

Buss, C., Wolf, O. T., Witt, J., & Hellhammer, D. H. (2004). Autobiographic memory impairment following acute cortisol administration. *Psychoneuroendocrinology, 29*(8), 1093–1096.

Cheung, S. S., D'eon, N. J., & Brooks, C. J. (2001). Breath-holding ability of offshore workers inadequate to ensure escape from ditched helicopters. *Aviation, Space, and Environmental Medicine, 72*(10), 912–918.

Cian, C., Koulmann, N., Barraud, P. A., Raphel, C., Jimenez, C., & Melin, B. (2000). Influences of variations in body hydration on cognitive function: effect of hyper-hydration, heat stress, and exercise-induced dehydration. *Journal of Psychophysiology, 14*(1), 29.

Civil Aviation Authority. (2014). *Safety review of offshore public transport helicopter operations in support of the exploitation of oil and gas CAP 1145.* Accessed from http://www.caa.co.uk/docs/33/CAP%201145%20Offshore%20helicopter%20review%20and%20annexes%2024214.pdf.

Cohen, R. M., Weingartner, H., Smallberg, S. A., Pickar, D., & Murphy, D. L. (1982). Effort and cognition in depression. *Archives of General Psychiatry, 39*(5), 593.

Cohen, S., & Wills, T. A. (1985). Stress, social support, and the buffering hypothesis. *Psychological Bulletin, 98*(2), 310.

Cowan, N. (1999). *An embedded-processes model of working memory. Models of working memory: Mechanisms of active maintenance and executive control,* 62–101.

Cunningham, W. F. (1978). Helicopter underwater escape trainer (9D5). In *AGARD Conference Proceedings, No. 255 (Operational helicopter aviation medicine)* (pp. 66-1–66-3).

Darke, S. (1988). Anxiety and working memory capacity. *Cognition and Emotion, 2*(2), 145–154.

de Quervain, D. J., Roozendaal, B., Nitsch, R. M., McGaugh, J. L., & Hock, C. (2000). Acute cortisone administration impairs retrieval of long-term declarative memory in humans. *Nature: Neuroscience, 3*, 313–314.

Deladisma, A. M., Gupta, M., Kotranza, A., Bittner, J. G., Imam, T., Swinson, D., et al. (2009). A pilot study to integrate an immersive virtual patient with a breast complaint and breast examination simulator into a surgery clerkship. *The American Journal of Surgery, 197*(1), 102–106.

Dickerson, S. S., & Kemeny, M. E. (2004). Acute stressors and cortisol responses: a theoretical integration and synthesis of laboratory research. *Psychological Bulletin, 130*(3), 355.

Dietrich, A., & Sparling, P. B. (2004). Endurance exercise selectively impairs prefrontal-dependent cognition. *Brain and Cognition, 55*(3), 516–524.

Dollins, A. B., Krock, L. P., Storm, W. F., Wurtman, R. J., & Lieberman, H. R. (1995). L-tyrosine ameliorates some effects of lower body negative pressure stress. *Physiological Behaviour, 57*, 223–230.

Eilam, D., Izhar, R., & Mort, J. (2011). Threat detection: behavioral practices in animals and humans. *Neuroscience & Biobehavioral Reviews, 35*(4), 999–1006.

Entin, E. E., & Serfaty, D. (1990). *Information gathering and decision making under stress.* NTIS HC A05. MF A01. Technical Report Number ADA218233.

Eysenck, M. W., & Calvo, M. G. (1992). Anxiety and performance: the processing efficiency theory. *Cognition & Emotion, 6*(6), 409–434.

Fosnaugh, J., Geers, A. L., & Wellman, J. A. (2009). Giving off a rosy glow: the manipulation of an optimistic orientation. *The Journal of Social Psychology, 149*(3), 349–364.

Funahashi, S. (2001). Neuronal mechanisms of executive control by the prefrontal cortex. *Neuroscience Research, 39*, 147–165.

Gerin, W., Pieper, C., Levy, R., & Pickering, T. G. (1992). Social support in social interaction: a moderator of cardiovascular reactivity. *Psychosomatic Medicine, 54*(3), 324–336.

Gottlieb, B. H., & Bergen, A. E. (2010). Social support concepts and measures. *Journal of Psychosomatic Research, 69*(5), 511–520.

Greene, M., Perry, R., & Lindell, M. (1981). The March 1980 eruptions of Mt. St. Helens: citizen perceptions of volcano threat. *Disasters, 5*(1), 49–66.

Grothmann, T., & Patt, A. (2005). Adaptive capacity and human cognition: the process of individual adaptation to climate change. *Global Environmental Change, 15*(3), 199–213.

Hartley, L. R., & Adams, R. G. (1974). Effect of noise on the Stroop test. *Journal of Experimental Psychology, 102*(1), 62.
Heinrichs, M., Baumgartner, T., Kirschbaum, C., & Ehlert, U. (2003). Social support and oxytocin interact to suppress cortisol and subjective responses to psychosocial stress. *Biological Psychiatry, 54*(12), 1389–1398.
Hytten, K. (1989). Helicopter crash in water: effects of simulator escape training. *Acta Psychiatrica Scandinavica, 80*(s355), 73–78.
Hytten, K., Jensen, A., & Vaernes, R. (1989). Effects of underwater escape training: a psychophysiological study. *Aviation, Space, and Environmental Medicine, 60*(5), 460–464.
Janis, I. L., & Mann, L. (1977). *Decision making: A psychological analysis of conflict, choice, and commitment.* New York: Free Press.
Jarvis, M. J. (1993). Does caffeine intake enhance absolute levels of cognitive performance? *Psychopharmacology, 110*(1–2), 45–52.
Kamarck, T. W., Manuck, S. B., & Jennings, J. R. (1990). Social support reduces cardiovascular reactivity to psychological challenge: a laboratory model. *Psychosomatic Medicine, 52*(1), 42–58.
Kirschbaum, C., Prüssner, J. C., Stone, A. A., Federenko, I., Gaab, J., Lintz, D., et al. (1995). Persistent high cortisol responses to repeated psychological stress in a subpopulation of healthy men. *Psychosomatic Medicine, 57*(5), 468–474.
Kirschbaum, C., Wolf, O. T., May, M., Wippich, W., & Hellhammer, D. H. (1996). Stress- and treatment-induced elevations of cortisol levels associated with impaired declarative memory in healthy adults. *Life Sciences, 58*(17), 1475–1483.
Kivimäki, M., Vahtera, J., Elovainio, M., Helenius, H., Singh-Manoux, A., & Pentti, J. (2005). Optimism and pessimism as predictors of change in health after death or onset of severe illness in family. *Health Psychology, 24*(4), 413.
Lavie, N. (1995). Perceptual load as a necessary condition for selective attention. *Journal of Experimental Psychology: Human Perception and Performance, 21*, 451–468.
Lazarus, R. S. (1966). *Psychological stress and the coping process.* New York, NY, US: McGraw-Hill.
Leach, J. (1994). *Survival psychology.* UK: Macmillan Publishing.
Leach, J. (2004). Why people 'freeze' in an emergency: temporal and cognitive constraints on survival responses. *Aviation Space Environmental Medicine, 75*, 539–542.
Leach, J. (2005). Cognitive paralysis in an emergency: the role of the supervisory attentional system. *Aviation Space Environmental Medicine, 76*, 134–136.
Leach, J., & Griffith, R. (2008). Restrictions in working memory capacity during parachuting: a possible cause of 'no pull' fatalities. *Applied Cognitive Psychology, 22*(2), 147–157.
Lieberman, H. R., Tharion, W. J., Shukitt-Hale, B., Speckman, K. L., & Tulley, R. (2002). Effects of caffeine, sleep loss, and stress on cognitive performance and mood during US Navy SEAL training. *Psychopharmacology, 164*(3), 250–261.
Lorist, M. M., Boksem, M. A., & Ridderinkhof, K. R. (2005). Impaired cognitive control and reduced cingulate activity during mental fatigue. *Cognitive Brain Research, 24*(2), 199–205.
Lupien, S. J., & McEwen, B. S. (1997). The acute effects of corticosteroids on cognition: integration of animal and human model studies. *Brain Research Reviews, 24*(1), 1–27.
Lupien, S. J., Gillin, C. J., & Hauger, R. L. (1999). Working memory is more sensitive than declarative memory to the acute effects of corticosteroids: A dose-response study in humans. *Behavioral Neuroscience, 113*(3), 420.
Maguire, E. A., Frackowiak, R. S., & Frith, C. D. (1997). Recalling routes around London: activation of the right hippocampus in taxi drivers. *The Journal of Neuroscience, 17*(18), 7103–7110.

Mäkinen, T. M. (2007). Human cold exposure, adaptation, and performance in high latitude environments. *American Journal of Human Biology, 19*(2), 155–164.

Miller, D. B., & O'Callaghan, J. P. (2002). Neuroendocrine aspects of the response to stress. *Metabolism, 51*(6), 5–10.

Milner, B. (1982). Some cognitive effects of frontal-lobe lesions in man. *Philosophical Transactions of the Royal Society of London B: Biological Sciences, 298*(1089), 211–226.

Miyake, A., & Shah, P. (Eds.). (1999). *Models of working memory: Mechanisms of active maintenance and executive control.* Cambridge: Cambridge University Press.

Morgan, C. A., III, Wang, S., Southwick, S. M., Rasmusson, A., Hazlett, G., Hauger, R. L., et al. (2000). Plasma neuropeptide-Y concentrations in humans exposed to military survival training. *Biological Psychiatry, 47*(10), 902–909.

Morgan, C. A., Hazlett, G., Doran, A., Garrett, S., Hoyt, G., Thomas, P., et al. (2004). Accuracy of eyewitness memory for persons encountered during exposure to highly intense stress. *International Journal of Law and Psychiatry, 27*(3), 265–279.

Newcomer, J. W., Selke, G., Melson, A. K., Hershey, T., Craft, S., Richards, K., et al. (1999). Decreased memory performance in healthy humans induced by stress-level cortisol treatment. *Archives of General Psychiatry, 56*(6), 527.

Nezafat, R., Shadmehr, R., & Holcomb, H. H. (2001). Long-term adaptation to dynamics of reaching movements: a PET study. *Experimental Brain Research, 140*(1), 66–76.

Ozel, F. (2001). Time pressure and stress as a factor during emergency egress. *Safety Science, 38*(2), 95–107.

Payne, J. W., Bettman, J. R., & Johnson, E. J. (1993). *The adaptive decision maker.* Cambridge: Cambridge University Press.

Pinquart, M., & Duberstein, P. R. (2010). Depression and cancer mortality: a meta-analysis. *Psychological Medicine, 40*(11), 1797.

Pires, T. T. (2005). An approach for modeling human cognitive behavior in evacuation models. *Fire Safety Journal, 40*(2), 177–189.

Porter, H., & Leach, J. (2010). Executive dysfunction in a survival environment. *Applied Cognitive Psychology, 24*(1), 41–66.

Qin, S., Hermans, E. J., van Marle, H. J., Luo, J., & Fernández, G. (2009). Acute psychological stress reduces working memory-related activity in the dorsolateral prefrontal cortex. *Biological Psychiatry, 66*(1), 25–32.

Ravindran, A. V., Matheson, K., Griffiths, J., Merali, Z., & Anisman, H. (2002). Stress, coping, uplifts, and quality of life in subtypes of depression: a conceptual frame and emerging data. *Journal of Affective Disorders, 71*(1), 121–130.

Ren, X. S., Skinner, K., Lee, A., & Kazis, L. (1999). Social support, social selection and self-assessed health status: results from the veterans health study in the United States. *Social Science & Medicine, 48*(12), 1721–1734.

Reznek, M., Harter, P., & Krummel, T. (2002). Virtual reality and simulation: training the future emergency physician. *Academic Emergency Medicine, 9*(1), 78–87.

Robinson, S. J., Leach, J., & Bridges, N. (2010). The role of social support in mediating subjective and physiological responses to an acute naturalistic stressor – helicopter underwater evacuation Training. In *Proceedings of the BPS Psychobiology Section.*

Robinson, S. J., Leach, J., Owen-Lynch, P. J., & Sünram-Lea, S. I. (2013). Stress reactivity and cognitive performance in a simulated firefighting emergency. *Aviation, Space, and Environmental Medicine, 84*(6), 592–599.

Robinson, S. J., Leach, J., & Walsh, J. (2009). The impact on executive functioning of exposure to an acute naturalistic stressor. In *Proceedings of the BPS Psychobiology Section.*

Robinson, S. J., Sünram-Lea, S., Leach, J., & Owen-Lynch, P. J. (2008). The effects of exposure to an acute naturalistic stressor on working memory, state anxiety and salivary cortisol concentration. *International Journal on the Biology of Stress, 11*(2), 115–124.

Rosen, V. M., & Engle, R. W. (1998). Working memory capacity and suppression. *Journal of Memory and Language, 39*(3), 418–436.

Roy, M. P., Steptoe, A., & Kirschbaum, C. (1998). Life events and social support as moderators of individual differences in cardiovascular and cortisol reactivity. *Journal of Personality and Social Psychology, 75*(5), 1273.

Ryack, B. L., Luria, S. M., & Smith, P. F. (1986). Surviving helicopter crashes at sea: a review of studies of underwater egress from helicopters. *Aviation, Space, and Environmental Medicine, 57*, 603–609.

Scheier, M. E., & Carver, C. S. (1987). Dispositional optimism and physical well being: the influence of generalized outcome expectancies on health. *Journal of Personality, 55*(2), 169–210.

Schuh, K. J., & Griffiths, R. R. (1997). Caffeine reinforcement: the role of withdrawal. *Psychopharmacology, 130*(4), 320–326.

Scoville, W. B., & Milner, B. (1957). Loss of recent memory after bilateral hippocampal lesions. *Journal of Neurology, Neurosurgery, and Psychiatry, 20*(1), 11.

Segerstrom, S. C., Taylor, S. E., Kemeny, M. E., & Fahey, J. L. (1998). Optimism is associated with mood, coping, and immune change in response to stress. *Journal of Personality and Social Psychology, 74*(6), 1646.

Seligman, M. E. (2011). *Learned optimism: How to change your mind and your life.* New York: Random House Digital, Inc.

Selye, H. (1974). *Stress without distress.* New York: New American Library.

Sharkey, T. J., & McCauley, M. E. (1992). Does a motion base prevent simulator sickness. In *Proceedings of the AIAA/AHS flight simulation technologies conference* (pp. 21–28).

Shimamura, A. P., Janowsky, J. S., & Squire, L. R. (1990). Memory for the temporal order of events in patients with frontal lobe lesions and amnesic patients. *Neuropsychologia, 28*(8), 803–813.

Smith, D. L., Petruzzello, S. J., Chludzinski, M. A., Reed, J. J., & Woods, J. A. (2001). Effects of strenuous live-fire fighting drills on hematological, blood chemistry and psychological measures. *Journal of Thermal Biology, 26*(4–5), 375–379.

Smith, M. A., Riby, L. M., Eekelen, J. A. M., & Foster, J. K. (2011). Glucose enhancement of human memory: a comprehensive research review of the glucose memory facilitation effect. *Neuroscience & Biobehavioral Reviews, 35*(3), 770–783.

Sünram-Lea, S. I., Owen-Lynch, J., Robinson, S. J., Jones, E., & Hu, H. (2012). The effect of energy drinks on cortisol levels, cognition and mood during a fire-fighting exercise. *Psychopharmacology, 219*(1), 83–97.

Taber, M. J. (2013). Crash attenuating seats: effects on helicopter underwater escape performance. *Safety Science, 57*, 179–186.

Taber, M. J. (2014). Simulator fidelity and contextual interference in helicopter underwater egress training – an analysis of training and retention of egress skills. *Safety Science, 62*, 271–278.

Taber, M. J., Kozey, J. W., & McCabe, J. (2012). *Investigation of emergency breathing apparatus skill set knowledge transfer between helicopter underwater escape training simulators [Online].* Nova Scotia, Canada: Survival Systems Training Ltd. Accessed from http://www.sstl.com/flow/uploads/HUEBAabstract.pdf on the 20/06/14.

Taber, M. J., & McCabe, J. (2006). Helicopter ditching: time of crash and survivability. *SAFE Journal, 34*(1), 5–10.

Taber, M. J., & McCabe, J. (2009). The effect of emergency breathing systems during helicopter underwater escape training for land force troops. *Safety Science, 47*(8), 1129–1138.

Taber, M. J., & McGarr, G. W. (2013). Confidence in future helicopter underwater egress performance. *Safety Science, 60*, 169–175.

Thomas, J. R., Lockwood, P. A., Singh, A., & Deuster, P. A. (1999). Tyrosine improves working memory in a multitasking environment. *Pharmacology, Biochemistry and Behaviour, 64*(3), 495–500.

Tipton, M. J., Gibbs, P., Brooks, C., Roiz de Sa, D., & Reilly, T. J. (2010). ECG during helicopter underwater escape training. *Aviation, Space, and Environmental Medicine, 81*(4), 399–404.

Tobin, D. L., Holroyd, K. A., Reynolds, R. V., & Wigal, J. K. (1989). The hierarchical factor structure of the coping strategies inventory. *Cognitive Therapy and Research, 13*(4), 343–361.

Tops, M., Van Der Pompe, G., Wijers, A. A., Den Boer, J. A., Meijman, T. F., et al. (2004). Free recall of pleasant words from recency positions is especially sensitive to acute administration of cortisol. *Psychoneuroendocrinology, 29*(3), 327–338.

Uchino, B. N. (2006). Social support and health: a review of physiological processes potentially underlying links to disease outcomes. *Journal of Behavioral Medicine, 29*(4), 377–387.

Weinstein, N. D. (1987). Unrealistic optimism about susceptibility to health problems: conclusions from a community-wide sample. *Journal of Behavioral Medicine, 10*(5), 481–500.

Wells, R. (2010). *Canada-Newfoundland and Labrador offshore helicopter safety inquiry* (Vol. 1). St. John's, NL: Canada-Newfoundland and Labrador Offshore Petroleum Board.

Wolf, O. T., Schommer, N. C., Hellhammer, D. H., Reischies, F. M., & Kirschbaum, C. (2002). Moderate psychosocial stress appears not to impair recall of words learned 4 weeks prior to stress exposure. *Stress, 5*(1), 59–64.

Xu, J., Mendrek, A., Cohen, M. S., Monterosso, J., Rodriguez, P., Simon, S. L., et al. (2005). Brain activity in cigarette smokers performing a working memory task: effect of smoking Abstinence. *Biological Psychiatry, 58*(2), 143–150.

Yesavage, J. A., Mumenthaler, M. S., Taylor, J. L., Friedman, L., O'Hara, R., Sheikh, J., et al. (2002). Donepezil and flight simulator performance: effects on retention of complex skills. *Neurology, 59*(1), 123–125.

CHAPTER 7

Use and Implications of Fidelity in Helicopter Underwater Egress Training

Michael J. Taber
Falck Safety Services Canada Inc., Dartmouth, NS, Canada

7.1 INTRODUCTION

The purpose of this chapter is to outline how different types of fidelity might influence the acquisition, transfer, and retention of helicopter underwater egress training (HUET) skills from training to real-world environments. In general terms, fidelity (explained in further detail below) is often used to describe how closely a simulated environment replicates the real world. Based on previous motor learning research (Adams, 1987; Anderson, 2008; Schmidt, 1975, 2003; Schmidt & Lee, 2011), the differences in current training environments are believed to contribute to situations in which individuals may not be able to appropriately transfer the physical skills practiced during a low-fidelity HUET program to an emergency situation (i.e., ditching) (Taber, 2014a). This belief is based on anecdotal evidence gathered from actual ditching events as well as assessments of generic-type underwater egress simulators (UESs) and benign environmental conditions used during training. From the perspective of high-fidelity simulation (HFS), this chapter is intended to provide HUET program facilitators and offshore operators with relevant knowledge about what type, how much, and when fidelity influences the ability to properly prepare individuals for an event that could require them to egress from an inverted and submerged position. Direction on the use of specific physical components is also provided in an effort to identify how HFS is important during skill acquisition. To aid the reader in understanding why it is important to consider fidelity in HUET programs, the information is contextualized by providing examples of survivor testimony from several egress events.

Handbook of Offshore Helicopter Transport Safety
ISBN 978-1-78242-187-0
http://dx.doi.org/10.1016/B978-1-78242-187-0.00007-9

7.2 CONTEXTUALIZATION OF FIDELITY

One of the first officially documented survivor accounts of a ditching event specifically addressing the difference between HUET program fidelity and a real environment can be found in a report of the North Sea Super Puma (G-TIGH) ditching/water impact in 1992 (Civil Aviation Authority (CAA), 1995). At approximately 1950 hours on the 14th of March 1992, a Super Puma helicopter carrying 17 passengers and crew to a nearby accommodation barge (Safe Supporter) ditched into the North Sea (Woolfson, Foster, & Beck, 1996). According to reports from survivors, the helicopter inverted and sank rapidly after impacting the surface of the water. There were 11 fatalities (five bodies recovered from the underwater wreckage and six from the surface). It was noted in the CAA (1995) report that, "Both [survivors] considered that the courses had been good value and had contributed to their survival, although the underwater escape simulator bore little or no resemblance to the passenger cabin of the real helicopter" (p. 43, brackets added). Similar comments can be found in the survivors' comments for the S92 ditching off the coast of Newfoundland and Labrador in 2009 and the Super Puma ditching off the coast of the Shetland Islands in 2013. The lone survivor from the S92 ditching reported that, "A couple days of controlled immersions in a pool every few years is not enough to allow anyone to develop the instinct of reaction that they need to have a chance of escaping a helicopter crash like Cougar 491" (Decker testimony, p. 85—www.oshsi.nl.ca/userfiles/files/HELN5.pdf). When asked to elaborate on his experience during the egress, he indicated that, "The chances of being able to escape from an overturned helicopter being on the inside seat would be next to impossible ... You would have to hold your breath and wait for the initial person who would be directly next to the window to get out and clear out of your way. Their feet are kicking. I just cannot see how this person would ever stand a chance" (Decker testimony, p. 90—www.oshsi.nl.ca/userfiles/files/HELN5.pdf). The survivor's comments appear to reflect the fact that this type of UES training was not in place prior to the event and that he had not had the opportunity to practice this type of skill. Had the underwater egress training included similar aspects known to exist in an actual helicopter, it is possible to speculate that his comments might have been different. Finally, one of the survivors from the Shetland Islands Super Puma ditching indicated that the exit in the actual helicopter was harder to open than it was during the training (Knass, 2013). In fact, the survivor commented, "I had hold of the tab on the window, I pulled but it came to bits. I was pushing the window out with my elbow but it wouldn't move—I punched it two or three times and it popped out" (Knass, 2013). This survivor also commented on the fact that others who had removed their seat belts while the water rushed in were floating around. Despite having successfully completed HUET prior to the event, these individuals apparently did not follow one of the most basic aspects of their training (i.e., ensuring a physical reference point before ever removing the seat belt).

Notwithstanding the comments that have been made about a lack of similarities between the training and the real-world environment, some ditching survivor accounts of past training are encouraging. For example, one military ditching survivor attests that, "Your training is so realistic that to appreciate the value of your techniques, one has to survive a real situation" (Survival Systems Limited—http://www.survivalsystemsgroup.com/testimonials.asp). Although somewhat generic in nature, numerous people have pointed out that the HUET program process was extremely beneficial to their survival (Ryack, Luria, & Smith, 1986) and there is no doubt that without the training, some of these individuals would probably not have survived. The link between training and survival has also been highlighted in several documented reports, and Cunningham (1978) clearly showed that there was a considerable difference in survival rates for those who had completed HUET (91.5%) compared to those who had not (66%).

Given the various details provided by survivors and information contained in accident reports regarding the level of fidelity included in the HUET program, it is important to consider what type of simulation is included in training, when it should be implemented, and how much realism is appropriate to ensure that the inclusion of HFS is beneficial in the event that the skills need to be used in a real ditching or water impact. Based on a review of the training environments, the CAA (2014) suggests that there is a clear need for realistic conditions during skill acquisition. This is also echoed by Commissioner Wells (2010) and is described as "greater fidelity" and "more realistic" in Recommendation 13 of the public inquiry report. Unfortunately, there is no clear delineation of how much HFS there should be or what this "realistic" environment should look like. Without a formalized process to achieve clearly defined objectives, HUET program providers are free to include as much or as little HFS as is believed to be necessary.

7.3 USE OF SIMULATED ENVIRONMENTS FOR TRAINING

It has been noted that the primary objectives for using simulated environments during training are to reduce cost, increase safety, and enhance the flexibility in the type of training scenario (Simmons, Lees, & Kimball, 1978).

From a safety perspective, the use of simulation is extremely valuable, as it provides individuals with the opportunity to test out and practice skills that would otherwise be too dangerous in a real setting. This is particularly true for learning the skills needed to egress from a capsized and flooded (possibly sinking) helicopter (Taber & McGarr, 2013). Unlike other skills training environments (e.g., learning to perform a non-life-threatening task), there is no "safe" way to train for underwater egress in the real world; therefore, simulation is the ideal option.

Unfortunately, the exact level and type of simulation have not been quantified for HUET skill acquisition. As an example of the variability in the training environments, Taber (2014a) suggests that, based on a Google search, "there are at least 33 different types of simulators being used to train individuals how to egress from a capsized aircraft" (p. 272). This variability in simulation standards creates a nonstandardized potential for various levels of skill training as well as a limitation in the ability to assess competency. Moreover, the various training environments are not necessarily used to increase the level of skilled practice. For example, numerous training facilities teach the same basic egress skills (regardless of level of fidelity) in both introductory and subsequent courses. Currently, there is no possibility for offshore personnel to complete an "advanced" underwater egress program that requires increased complexity in the egress tasks (e.g., low-level light conditions or moving over a row of seats to egress while waiting for several others to escape). Without the possibility of learning a new (more advanced) set of related skills, there is no way of increasing an individual's understanding of the required tasks under different circumstances.

As there is no opportunity for advancement, it can be argued that if the level of fidelity in a simulated environment limits the assessment of HUET skill competency (i.e., no possibility to complete an aisle or cross-cabin egress—Cougar 491 survivor testimony), there exists the possibility that some individuals will not gain the practical experience that may be needed in an actual ditching. These limitations associated with the simulated environment naturally lead to a question of how international training providers and operators recognize HUET certification between offshore jurisdictions. This is particularly problematic if the helicopter(s) being used to transport the individuals to and from the offshore installation requires specific egress or evacuation tasks. Without a clear understanding of what type and how much fidelity is optimally beneficial for acquiring HUET skills, international standards will remain vague and open to interpretation.

7.4 PART-TASK VERSUS WHOLE-TASK TRAINING

Taber and McGarr (2013) point out that the inclusion of a whole-task (entire sequence of tasks) demonstration is important when considering how individuals develop a mental model of the performance of complex tasks (e.g., the use of emergency breathing systems; Chapter 8). Given the complexity of the egress tasks, it is not surprising that trainees' performance of underwater egress skills during their first HUET program shows a clear indication that a learning effect occurs. It has been shown that as the trainees perform subsequence egress trial (runs) in the UES, their performance progressively improves (Kozey, McCabe, & Jenkins, 2007; Taber, Kozey, & McCabe, 2012; Taber & McGarr, 2013). Based on data collected from HUET performance in Canada and the United States (unpublished data), it has been noted that the percentage of failed attempts to open a simulated push-out exit for the first time underwater is approximately threefold higher compared to final attempts just before gaining certification. This decrease in errors is seen despite the fact that the runs become progressively more difficult (Chapter 2). It appears that as the trainee gains more experience opening an exit under water, there is a decrease in the number of failed attempts. Kozey et al. (2007) clearly showed that having the opportunity to practice the skill set needed to open an exit under water (i.e., higher level of fidelity) significantly improves the ability to perform the same task six months later. Similarly, Taber et al. (2012) found that individuals who had the most practice (eight trials versus none) using a compressed-air breathing system while in the UES required the least amount of assistance from the HUET instructor/expert assessor during the retention testing session (30 days later). This research suggests that the performance of a whole-task demonstration is necessary to ensure that individuals have gained the desired skills, are capable of combining the part tasks into one complex whole task, and have retained the procedural knowledge beyond an immediate recall period (Lim, Reiser, & Olina, 2009).

7.5 PHYSICAL/FUNCTIONAL/COGNITIVE SIMULATION FIDELITY

This section addresses the types of fidelity that are believed to be most important in a HUET program. Historically, the use of HFS in training is not a new topic. For example, a discussion or conceptualization of including realistic aspects in learning environments can be traced back to the writing of Aristotle (see Hays & Singer, 1989, for a detailed overview of learning theory). From these early beginnings of learning theory, a variety of definitions for the term "fidelity" have been offered from domains such as aviation, marine, transportation, and medicine (Bradley, 2006). Hays (1981) writes, "training simulator fidelity is the degree of similarity between the simulator and the equipment that is simulated" (p. 2). Based on this particular definition, different types of fidelity such as psychological/cognitive, physical, functional, task, and operational have been proposed (Thomas, 2003, p. 91). The main difference between each proposed type of fidelity reflects the specific focus on realism in a particular area. For example, Hays (1981) identifies the distinction between physical and functional characteristics by discussing training effectiveness. In the case of a HUET program, physical fidelity refers to aspects of the whole environment (e.g., environmental effects such as wind, waves, rain, and lighting conditions and the UES effects such as the descent speed, roll rate, seat configuration, harness design, seat design, exit placement and type—to mention only a few) in which the training is carried out. Functional fidelity, on the other hand, refers to aspects of the exit (e.g., force required to open or rotate handle) or seat harness (e.g., direction of rotation or force necessary to open the release mechanism) procedures.

Based on the information presented below, there is a clear need to consider the influence of both physical and functional fidelity from a holistic framework. It is particularly important when developing a training program (Chapter 2) that is designed to prepare individuals for a helicopter ditching or water impact. It is also clear that distinguishing the difference between physical, functional, and other kinds of fidelity is critical to an overall understanding of how much reality is too much and how much is too little. The wrong combination of fidelity could result in an increased exposure to risk (excessive stress and/or anxiety) or lack of preparation for an actual event.

7.5.1 Physical Fidelity

The concept of physical fidelity is intended to highlight the influence of various levels of realism in the transfer of motor skill knowledge from a simulated environment to a real-world setting and was first proposed by Thorndike and Woodworth (1901). More recently, Hochmitz and Yuviler-Gavish (2011) describe physical fidelity as "the degree to which a simulated environment replicates the look, sound, and feel of the real-world environment" (p. 490). This definition suggests that as the simulated environment increases in its visual, tactile, and auditory resemblance to the actual environment, physical fidelity increases. It also suggests that a holistic consideration of more than just the functionality of an object should be included in the assessment of fidelity.

For HUET programs, the term physical fidelity is not meant to include only the environment that is immediately next to the person. Rather, in addition to the items within close proximity to the person, physical fidelity is meant to include simulated aspects of the entire experience (e.g., speed of descent, amount of impact force, speed of inversion, lighting, inclusion of wind/waves, number of other people egressing, inclusion of appropriate supplementary survival equipment, and requirement to complete postegress survival skills). A consideration of all of the additional components known to exist during a real event ensures that the entire simulation (not just the egress simulator) is designed to better prepare the individuals (Taber, 2014b).

A primary limitation associated with low physical fidelity during the initial acquisition of a skill is the fact that a mental schema created for that particular situation is dependent on the environmental cues necessary to initiate the sequence of tasks at some future date. If, for example, the simulated environment is substantially different from what would be expected in a real event (e.g., exit type/functionality/distance from seat, seat type and functionality, configuration of seats and exits, roll characteristic of the egress simulator), the skills that have been learned during training may not be sufficient to egress the actual ditching. Success in training does not always ensure success during a real event. Therefore, it is extremely important to ensure that individuals have the opportunity to learn a skill in a safe environment that has a high level of physical fidelity (Hays, Jacobs, Prince, & Salas, 1992; Taber, 2014a; Taber & McGarr, 2013).

Figure 7.1 represents one of the highest known levels of physical fidelity that currently exists for HUET simulators. From the figure it is clear that the gap between HFS and the actual aircraft is minimal. Based on the nearly exact approximation, transfer of skills from one environment to another should require only minimal cognitive resources without the need to extrapolate training knowledge to a novel time-constrained situation (i.e., a ditching).

7.5.1.1 Physical Fidelity Guidance

Given the limited information regarding how much physical fidelity is necessary to ensure optimal skill acquisition and retention, it is not surprising that internationally recognized HUET programs vary in the amount of realism used during training. A limited Google search of "underwater helicopter escape images" (Taber, 2010, p. 211, 2014a) showed that

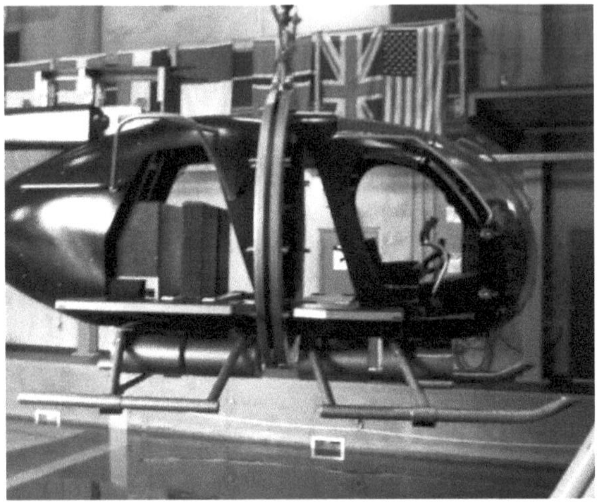

Figure 7.1 High level of physical fidelity simulation. *(Image from Survival Systems Limited, used with permission.)*

several new devices have been created and will probably continue to be created without the benefit of a formal guidance document. Unfortunately, without specific regulations or an agreed upon minimum standard from industry operators, simulator fidelity continues to range from a plastic chair bolted to a PVC frame to a full-scale UES that resembles the exact aircraft. Generic simulators may be sufficient for the development of basic skills such as when and how to adopt a brace position or when to take a breath of air before inversion; they do not provide an opportunity to practice specific skills that may be needed for a particular helicopter configuration (e.g., opening and exit from a crash-attenuated seat position, crossing an aisle, etc.).

From a physiological egress performance point of view, generic simulators have been shown to lack the capability to replicate disorientation (Cheung, Hofer, Brooks, & Gibbs, 2000). Cheung, Hofer, Brokks, and Gibbs (2000) clearly demonstrated that smaller simulators typically used in the shallow end of a pool do not produce a sufficient roll or descent rate to induce the physiological factors needed to create disorientation. Without this experience, it can be argued that individuals will not be prepared for the conditions most likely to occur in an underwater egress from a capsized and flooded helicopter that has descended to the water from altitude (Taber, 2014a).

7.5.1.2 Modular Egress Training Simulator (METS)™ Descent Rates

To consider the differences between an actual helicopter ditching/water descent rate and that of a UES that has a high level of physical fidelity in a training environment, an examination of the crane operating speeds has been explored. Kozey (2007) studied descent rate using three different crane speeds to lift/lower a modular egress training simulator (METS™). Based on previous research that investigated the influence of descent speed on disorientation (Cheung et al., 2000), the three descent rates were set at 32.8, 63, and 98.4 ft/min (10, 19.2, and 30 m/min). The three speeds represented both slower and faster descent rates than the 63 ft/min used at the time of testing. Kozey (2007) indicates that a METS™ in free fall (i.e., the cables or crane catastrophically fail) from 6.6 ft (2 m) would descend at a speed of 1230.3 ft/min (375 m/min) (obviously not used in a training environment). From the report it was clearly shown that even at the highest descent rate of 98.4 ft/min, the impact forces would be between 1.1 and 1.2 g, thus well within human tolerances (Kozey, 2007). To further qualify the impact forces, Kozey (2007) reported:

> *In terms of injury risk, the vertical decelerations are related to the loading on the humans seated in the METS. The decelerations were not affected by the different speeds and are comparable to the typical values seen in human walking and running studies (Winter, 1991). Therefore in terms of the vertical decelerations, the use of the 30.0 m/min speed should provide no greater risk than the current international standard of 19.2 m/min (p. 5).*

Free fall of a METS™ would be an extreme condition that would never be used in training; however, a Safety Alert published by NUTEC-MTC (2001) describes an incident in which the wire failed during underwater egress training. When the cable failed, the UES free fell to the water and sank to the bottom of the pool with two trainees inside. Both trainees egressed without incident and without injury (NUTEC-MTC, 2001). Similar events have occurred in Australia, The Netherlands, and Scotland (personal communication with Survival Systems Limited founder). In these cases, trainees and instructors were inside the simulator when it free fell to the water and no injuries were reported. Therefore, even with a predicted descent rate similar to those reported by Kozey (2007) and approximately half of the descent rate reported in the Cougar 491 ditching (TSBC, 2009), there were no injuries in the training environments.

The descent rates are provided as a further example of various levels of physical fidelity that could potentially influence the transfer of skills from the simulated environment to an actual event. Given the possibility of injury in a UES free fall event it is not likely (nor recommended) that a training provider would subject individuals to the risk. The information does, however, suggest that a close approximation of an actual ditching can occur in a training environment without injuries.

7.5.2 Transfer of Skill

From a practical standpoint, Baldwin and Ford (1988) suggest that increasing the physical fidelity of a simulation to be as close as possible to real-world conditions greatly increases the transfer of skills from one environment to the other. Thomas (2003, p. 91) suggests that physical fidelity is the "visual, kinaesthetic and spatial similarities between the simulated and real systems." This definition (as well as those listed above) is included here to highlight how fidelity (particularly physical fidelity) plays a considerable role in training motor skills. In contrast to procedural knowledge (physically demonstrate), physical fidelity is important in the development of declarative knowledge (verbally explain); however, the extent of similarity may not need to be precise.

The term "muscle memory" is often used to describe the process of developing a motor plan for a particular skill (Hossein, Fallah, & Raza, 2012; Shusterman, 2011). The term implies that once a skill has been learned or overlearned (practice beyond initial mastery—automaticity), it becomes second nature and can be performed automatically. From a physiological perspective, a muscle cannot "remember" how to perform a task (Muscolino, 2006). However, the term suggests that if the task is repeated enough times, it will be so ingrained in a person's brain that the person no longer needs to think about the steps (i.e., the task can be completed subconsciously). However, despite the inclination to believe that rote learning (memorization) is the best course of action, it is not the repetitive process of performing a skill over and over again that strengthens the neural linkages in the motor cortex (Shea & Morgan, 1979). Rather, it is the combined integration of associated factors and variability during each subsequent practice session that allows individuals to create different links for the same task (Wright et al., 2015). Studies in skill acquisition and contextual interference (see below) have repeatedly shown that practicing the same set of tasks in the exact same pattern does not produce a high level of skill retention (Porter & Magill, 2010; Schmidt & Bjork, 1992; Shea & Morgan, 1979; Stokes, Lai, Holtz, Rigsbee, & Cherrick, 2008). At a basic level, learning HUET skills by repeating the same process over and over might be effective to a certain extent; however, the mental schema is only ever challenged immediately after learning the particular tasks. Based on this type of training model, it is not surprising that rapid degradation of the HUET skills can be seen in the results of studies from Mills and Muir (1999), Kozey et al. (2007), and Taber et al. (2012).

Skill retention is discussed in more detail below; however, at this point it is important to note that environmental cues (i.e., physical fidelity) presented during the encoding of information (i.e., during skill acquisition) often aid in the recall process. For example, if the initial acquisition of a skill is carried out in a training environment that includes visual, tactile, auditory, and/or olfactory stimuli that are known to be present in the real-world setting, the transfer of skill performance from one situation to the other should be superior (Rose et al., 2000). Coleshaw (2006) has purported that although increased realism is positively correlated with increases in stress (see also Coleshaw, 2013, Chapters 6 and 8), HUET trainees have indicated that their confidence in future performance increases (see also Taber & McGarr, 2013).

Taber (2014a) suggested that the lack of physical fidelity found in generic simulators requires individuals to create mental schema for situations of which they have no practical experience as a basic reference point. This lack of physical information (i.e., how far away is the secondary egress exit or what physical steps are required to egress over an auxiliary fuel cell) limits the completeness of the motor plan. Filling in the gaps of an egress plan while under extreme time constraints, limited tactile sensory feedback, and reduced visual cues is likely to increase the probability of performance errors and reduce the possibility of a successful egress. That is not to say that individuals are not capable of quickly putting together a plan that has proven to be successful on some occasions. However, the likelihood of success is diminished as further demands are placed on an individual under these conditions.

Taber (2014a) further indicates that the replication of cabin interiors such as the placement or type of emergency exits in relation to the location and type of seats is important in the development of a relevant mental model. The sole survivor of the Cougar 491 event made this limitation plainly clear in his statement that he had not been given the opportunity to practice egressing from an aisle seat. Research has also shown that placement of the seat in relation to the exit influences the maximal jettison forces that can be generated by a person trying to egress (Taber & Sweeney, 2014). Training programs that physically and mentally prepare individuals for worst-case scenarios (Chapter 6) may be more demanding, but can be

Figure 7.2 Example of a generic HUET simulator *(Image from Taber (2014a), used with permission.)*

highly beneficial in providing a safe environment in which individuals can learn how to manage aspects of a real-world event.

Finally, and of the greatest concern, the skills required to egress a generic UES (Figure 7.2) with a high level of accuracy create a belief that an individual will be able to generalize the information gained in a low-fidelity environment to a different situation (e.g., different emergency exits, sitting in an aisle versus a window, or one row back from an exit) (Taber, 2014a). This misleading belief assumes that because the individual performed well in the generic environment, he or she will be capable of combining all of the necessary skills needed to egress despite the slightly different, more stressful, and unfamiliar environment.

7.5.3 Crash-Attenuating Seats

Incorporating a higher level of physical fidelity in the UES allows for the inclusion of training components such as crash-attenuating seats, possible snag hazards, and specific techniques for cross-aisle or cross-cabin egress. For example, Taber (2013) reported that there is a significant difference (requires more time and is rated as more difficult) between egressing from a standard seat compared to egressing from a crash-attenuating seat that has been lowered to simulate heavy impact forces (four seats bottomed out in the Cougar 491 event—TSB, 2010). Figure 7.3 shows the use of crash-attenuating seats during a HUET program in Canada. From the image, it is possible to identify specific aspects of the egress sequence that would not be

Figure 7.3 Simulated crash-attenuating seat placed next to an auxiliary fuel cell.

possible to practice while in a generic simulator. Of particular interest is the fact that the individual would not be capable of reaching the push-out exit while strapped into a seat. Without having the opportunity to practice this in a controlled environment, it is likely that the individual would disconnect the seat harness before locating a reference point on the back of the seat.

Before moving on to a discussion of functional and cognitive fidelity it is important to discuss the physical component in regard to limitations. As mentioned, fidelity can range from a basic to an advanced replication. The range from low to high physical fidelity represents the portion of the continuum that can be safely replicated in training (Figure 7.4). There is, however, a point at which adding more realism is no longer safe and adds no further benefits (identified as an extreme level). Once this point has been reached (end of safe zone), further risk mitigation techniques are unlikely to bring the simulated environmental components back to the point at which the training would be safe. Understanding this limitation as well as the ability to clearly define that point is critical to the future of HUET programs.

Figure 7.4 Hypothetical physical fidelity continuum.

7.5.4 Functional Fidelity

Functional fidelity is a component of physical fidelity and refers to the level of similarity between the task requirements in the simulated space and the real world. For example, if the simulated seat harness requires the same actions to release as those in the real helicopter or the exits require the same force and operational tasks to open/jettison as those required in the real helicopter, it could be said that there is a high level of functional fidelity. If, however, there is a low level of functional fidelity (i.e., no requirement to pull a release tab on a push-out exit or removal of a witness plate on a mechanical handle) in the simulated environment, there should be no expectation that this skill would transfer to the real-world environment.

Creating a high level of functional fidelity does not necessarily require a similarly high level of physical fidelity. For example, a UES environment could have an extremely low level of physical fidelity (i.e., no wind, wave, or rain; no cockpit; dissimilar seats and interior configuration) while having a high level of functional fidelity in regard to the seat belt and exit mechanisms. The need for high levels of physical fidelity might not be necessary for simplistic tasks that do not affect overall performance related to specific skill sequences (Summers, 1996); however, research has shown that higher realism produces better performance outcomes (Hochmitz & Yuviler-Gavish, 2011). Hochmitz and Yuviler-Gavish (2011) indicate "transfer from the first task [the simulation] to the second, real-world task occurs most efficiently when the two tasks have identical component activities. The more elements that are shared between the two tasks, the better the transfer" (p. 490).

To further the discussion of whole-task versus part-task training, it is important to consider how functional fidelity could be used to fully prepare an individual for the entire set of tasks necessary to survive a ditching situation (see Chapter 2 *vignette*). Given that functional fidelity focuses on one component of the egress task, it should be considered a part-task training environment (Gerathewohl, 1969). Regardless of how realistic the seat-belt release mechanism might be, disconnecting from the seat is only one aspect of the entire process. In contrast, a high level of physical fidelity within the UES environment incorporates a high level of functional fidelity for all of the subcomponents of the egress.

7.5.5 Cognitive Fidelity

In contrast to physical or functional fidelity, cognitive fidelity has been described as the degree to which a simulation replicates psychological and cognitive factors such as stress, anxiety, situation awareness, and decision-making requirements found in a real-world environment (Allen, Hays, & Buffordi, 1986; Hochmitz & Yuviler-Gavish, 2011; Kaiser & Schroeder, 2003, p. 435). Taber (2014b) notes that a UES training environment would have a high level of cognitive fidelity if trainees show an increase in sympathetic response (Kreibig, 2010) similar to that expected to be present in a real ditching. Not surprisingly, the egress skills necessary to successfully complete a HUET program have anecdotally been described as one of the hardest things that trainees have ever had to attempt.

Physiologically, Tipton, Gibbs, Roiz de Sa, and Reilly (2010) showed that UES training increased heart rate, Coleshaw (2006) reported an increase in cortisol (often called the stress hormone), and Robinson, Sunram-Lea, Leach, and Owen-Lynch (2008) indicated that HUET participants had higher state anxiety levels related to the anticipation of completing the underwater egress skill set (see Chapter 6). These physiological changes may be considered an indication of higher UES program cognitive fidelity and may be viewed by the individual as either a threat or a challenge depending on the person's perspective. For example, some trainees report that the training environment and required skill set demonstration is a great opportunity to practice or challenge themselves in a controlled environment. For these individuals, making a mistake during training is often considered a good learning event that can be corrected on the next egress trial, and it is unlikely that the training environment will be sufficient to elicit a high level of sympathetic nervous system responses. This threat versus challenge view of training plays a significant role in the perceived level of stress (Lazurus & Folkman, 1984, p. 117).

Cognitive fidelity in training is useful in that it helps prepare individuals to make decisions in stressful situations (Bell, Szczepkowski, Santarelli, & Schlachter, 2005). Cognitive behavioral modification techniques such as stress inoculation training (Meichenbaum & Deffenbacher, 1988) require a high level of cognitive fidelity during simulations; however, they do not necessarily require a high level of physical or functional fidelity to be effective in having individuals imagine the real-world environment conditions. Therefore, using a combination of high physical, functional, and cognitive fidelity during the acquisition of critical skills ensures that an individual can imagine how to complete the tasks in a stressful environment while actually having the experience of being able to perform the necessary tasks. Based on past ditching event information, and previous research, it can be argued that to properly prepare an individual or team to perform a complex task in a real-world event, demonstration of critical skills should be practiced in an HFS environment while ensuring that safety is maintained.

Despite the fact that multiple types of fidelity have been identified and researched, there appears to be a considerable gap in the collective knowledge about how much or when to introduce it into a HUET program. One of the only prescriptive aspects from previous research suggests that high levels of fidelity should not be introduced at a rate that makes it difficult for a novice to acquire the necessary skills. However, there is little information regarding an optimal combination of physical, functional, and cognitive fidelity. For example, it would probably be unwise to require a novice HUET trainee to begin the certification process on the most difficult egress exercise. If, however, there were an advanced program, it might be advisable to start testing where the introductory training left off.

7.6 CONTEXTUAL INTERFERENCE IN TRAINING

Contextual interference (CI) was first proposed by Shea and Morgan (1979) to describe a process of learning that did not follow a traditional rote memorization of tasks. In their work, it was suggested that during early learning, high variability in required task performance facilitates skill acquisition, retention, and transfer as well as improving an individual's capability to adapt to changing conditions (Shea & Morgan, 1979; Stokes et al., 2008). Taber (2014) highlights the fact that previous motor skills training research shows superior retention of skills if the initial acquisition includes CI (Barreiros, Figueiredo, & Godinho, 2007; Lee & Simon, 2004, p. 29; Porter & Magill, 2010).

Aside from cautioning about too much information too soon, CI has been proposed as a beneficial approach to introduce varying levels of fidelity and difficulty into training (Battig, 1979; Rose & Cristina, 2006; Schmidt & Bjork, 1992). CI research has been used in both simple and complex motor task development within domains such as firefighting (Schmidt & Wrisberg, 2008) and sport (Pauwels, Swinnen, & Beets, 2014). CI has also been called perturbation training (Cooke, Gorman, Myers, & Duran, 2013) and stress inoculation training (Meichenbaum & Deffenbacher, 1988). The concept of stress inoculation training is similar to the idea of vaccinating individuals against common forms of influenza. Typically, stress inoculation training is designed to incrementally increase the level of difficulty in training scenarios to the point at which individuals experience ever-increasing amounts of stress in a controlled manner. As the level of difficulty increases, individuals eventually reach a point at which perceived demands outweigh the available resources and stress increases.

Including CI into a HUET program can be accomplished by changing seat positions (normal or lowered to a position that represents a crash-attenuating seat), by positioning the individual on different sides of the egress simulator (Chapter 2), or by changing the type of exit to be jettisoned (Taber, 2013). By randomly presenting these task requirements, it is suggested that individuals will be capable of recalling the necessary skills better than if they were training to do the exact same tasks in the exact same sequence from the exact same seat. It is purported that increase in performance during retention is based on elaboration of the required skills during initial acquisition or through a reconstruction process (Wright et al., 2015). CI can also be increased by asking the trainees to perform HUET tasks in simulations that include additional tasks associated with environmental conditions (e.g., wind, waves, and rain).

7.7 PREDICTABILITY OF FUTURE PERFORMANCE

Predicting future performance based on a training environment is difficult even under ideal circumstances. The variability of factors (e.g., impact forces, inversion rate, number of individuals involved, seating configuration, exit type, sea conditions, time of day, amount of warning—to mention only a few) from one ditching to another is almost limitless. It is, however, possible to consider the factors that influence the transfer of skills from a training (practice) setting to a real event.

One of the primary goals of any HUET program is to better prepare individuals for a ditching event (Chapter 2). This is true regardless of the type of equipment used in the practical exercise sessions. However, when available resources to build or purchase a full-scale UES are limited, it is often suggested that "some training is better than nothing" despite the fact that physical or functional fidelity lacks similarity to the actual helicopter. This type of suggestion is made without empirical evidence and is considered common sense or best practice because it has always been done that way. However, negative transfer training and retroactive interference research (Wohldmann, Healy, & Bourne, 2008) would suggest that learning a skill to the point of automaticity (overlearning) in an environment that does not resemble the actual space in which the tasks will be carried out would impede future performance (Krakauer, Ghilardi, & Ghez, 1999; Pine, Krakauer, Gordon, & Ghez, 1996; Wohldmann, Healy, & Bourne, 2008; Panzer,

Wilde, & Shea, 2006). Specifically, it has been shown that simple tasks such as reaching and grasping items can be affected by the type of practice that is used during the initial skill acquisition process (Joiner & Smith, 2008). Given the importance of locating and activating an emergency exit, it is important to ensure that the practice environment is as close to the real environment as possible.

It is generally understood that training providers are trying to deliver the best possible course that can be done with available resources (Chapter 2). Unfortunately, this approach of "something is better than nothing" has been shown to limit the capability to predict future performance in a real event. Cheung et al. (2000) clearly showed that the level of disorientation (considered to be a major factor in the ability to egress) was significantly higher in the full-scale UES compared to a shallow-water egress chair that is typically used in the shallow end of the pool. Based on these findings, it is reasonable to assume that individuals who are trained in a generic-type device in the shallow end of the pool will experience some difficulties in transferring the practice skills in the event of an emergency.

When considering all of the possible influencing factors associated with physical, functional, and cognitive fidelity, it is important to include a discussion of how much reality is too much. As indicated in the hypothetical physical fidelity continuum (Figure 7.4 above), there is a point at which too much realism in a training environment creates a situation in which the risks outweigh the benefits. Figure 7.5 expands this concept further by displaying a theoretical distribution of HUET performance based on the level of physical, functional, and cognitive fidelity. From the figure it is possible to identify the point at which optimal HUET performance is achieved while maintaining an acceptable level of realism. To estimate this optimal level of fidelity in training, performance outcomes should be considered from an inverted-U perspective (classic Yerkes Dodson law, 1908). The curve depicted in Figure 7.5 remains theoretical, as it would be unsafe to test the extreme end of the physical fidelity continuum.

The levels along the *x*-axis (horizontal) represent the various amounts of physical, functional, and cognitive fidelity. Table 7.1 provides further information outlining the components in each type of fidelity. These categorizations represent only some of the possible components that should be considered while trying to identify the HUET zone of optimal performance.

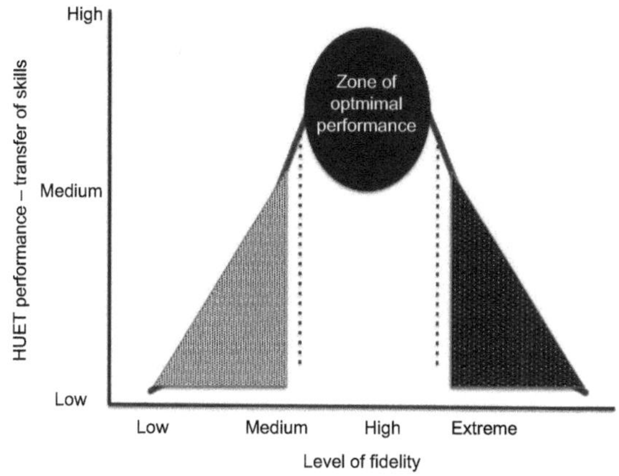

Figure 7.5 Theoretical curve of HUET performance transfer of skills based on level of fidelity. (*Note*: the area under the curve from low to medium fidelity would produce only minimal transfer of skills. While the extreme section could initially produce higher levels of transfer, it is likely that this performance outcome would rapidly drop off and the risks of injury are so great that it is not recommended for training.)

7.7.1 Low-Level Fidelity HUET Simulation

The label "low" in Table 7.1 represents a device and environment that do not replicate conditions that would be present in an actual event (e.g., multiple personnel, a descent from altitude). Moreover, generic exits (if any) used on low-level fidelity devices typically have no direct link to the functionality of the actual helicopter. This reduction in physical and functional fidelity will rarely induce the same level of stress/anxiety as a full-scale UES.

7.7.2 Medium-Level Fidelity HUET Simulation

The second level of categorization (medium) represents the initial integration of some physical and functional fidelity; however, it does not include all aspects of an actual event. For example, a medium-level fidelity simulation might include multiple seats and exits, but lacks the ability to include a reasonable descent or inversion rate to induce disorientation. This level of simulation is also highlighted by having limited similarities to the actual functionality of the helicopter in which most of the passengers travel. The use of generic exit styles does not allow the trainees the opportunity to practice skill sets that would be required in an actual ditching. Generic exits are also likely to be easier to open than those in the actual helicopter, thus resulting in high pass rates during training. Taber and McGarr (2013) discuss

Table 7.1 HUET Fidelity Level Categorization

Fidelity Type	Level Categorization			
	Low	**Medium**	**High**	**Extreme**
Physical	• Single-seat training device that is used in the shallow end of the pool • No environmental conditions (wind, rain, waves) • No descent rate • Inconsistent roll rate	• Multiple-seat training device used in the deep end of a pool • No environmental conditions • Generic exits • Slow descent rate (does not induce disorientation) • Roughly consistent roll rate	• Multiple-seat training device used in the deep end of a pool • Use of environmental conditions • Descent rate of at least 19.2 m/min (Kozey, 2007) • Consistent/controllable roll rate	• Multiple- or single-seat training device used in the deep end of a pool • Excessive use of environmental conditions (hazardous for staff and trainees) • Rapid descent rates that induce impact forces—increased risk of injury
Functional	• Generic exit • Generic seat belts	• Generic exits • Generic seat belts	• Specialized helicopter-specific exits • Helicopter-specific seat belts	• Specialized or generic exits (difficult to open—possibility of entrapment)
Cognitive	• Minimal increase in stress/anxiety	• Increased stress/anxiety	• Increased stress/anxiety	• Excessive increase in stress/anxiety to the point that trainee cannot continue the program

the limitations of using pass rates as an indication of successful competency attainment. Being able to open a simulated push-out exit that bears no resemblance to the one in the helicopter does not automatically ensure that the individual will be able to execute the necessary tasks in an actual emergency. Stress/anxiety levels for a medium level of fidelity are often higher than they would be for a low-level simulator because being strapped into a larger device that is lifted, lowered, and then inverted in the deep end of a pool increases the perception of risk. Although there may be a high level of perceived risk, the actual risk level for training in a certified/ industry-audited medium-fidelity UES is similar to or exactly the same as that of a low-fidelity environment (i.e., possibility of aspirating water while below the surface). In both cases, the risks are mitigated by having highly trained personnel close at hand (i.e., able to physically touch the person egressing the simulator) and by having a well-established emergency response protocol in place.

7.7.3 High-Level Fidelity HUET Simulation

A high-fidelity environment represents the combination of several aspects from physical, functional, and cognitive factors. Similar to the medium-level simulation, the HFS is critical to developing a specific mental model that can be directly transferred to the real world. Having the opportunity to perform/practice the entire set of tasks (whole-task) in an environment that includes the same type of exits positioned next to the same type of seats that are equipped with the same type of seat belts will undoubtedly provide an individual with an advantage. Enhancing this experience with environmental conditions such as wind, rain, and waves will ensure that issues associated with both the egress and the postegress survival and rescue phases have been explored (Taber, 2014b). Despite the obvious benefits, it is important for training organizations to manage the point at which actual training risks become too high.

7.7.4 Extreme-Level Fidelity HUET Simulation

At the extreme level of fidelity, the simulated environment could potentially injure the trainees (both physically and psychologically). For example, simulating the descent rate of a UES that could potentially cause impact injuries or unconsciousness would be extremely realistic; however, it would also be extremely dangerous. Training that fits into this category would provide the best transfer of knowledge to a real-world environment, but it is unlikely that anyone who survived would be willing to complete a second egress exercise. Identifying the point at which any of the HFS aspects becomes extreme is critical to ensuring a safe yet effective training environment in which to learn the necessary skills.

7.8 SUMMARY

The level of physical fidelity used during initial skill acquisition depends primarily on the complexity of the task. If, for instance, the task is simple in nature, the level of physical fidelity does not necessarily need to be very high. If, however, the skill requires precise movements that can be influenced by environmental conditions and is constrained by sequential integration of movements, physical fidelity is more critical than functional fidelity. Erecting the canopy of a specific aviation life raft (e.g., RFD) while under conditions that simulate waves, rain, and wind is a relevant example of how physical fidelity could potentially influence future performance of postegress skills. Completing similar tasks with a generic life raft without environmental conditions (low physical fidelity) is unlikely to prepare an individual for the performance of the tasks with a completely different canopy deployment system.

The same would be true for any number of tasks such as the jettisoning of an exit, release of a seat belt, activation of a personal locator beacon, deployment of a splash shield, or activation of a strobe light. Ideally, an individual is trained to complete the helicopter ditching and survival tasks in a controlled setting with the exact same equipment he or she will be using in the real event. Given the fact that humans rely significantly on visual cues to aid in skill acquisition, it is reasonable to assume that initial high physical fidelity simulation training should be conducted in a fully illuminated environment (Khan & Franks, 2004; Robertson, Tormos, Maeda, & Pascual-Leone, 2001). However, once the skill has been acquired, it should then be performed in an environment similar to that which could be expected in the real environment. Therefore if there is a potential for limited visibility (low-level light—night or fog), assessment of performance should be carried out under those conditions as long as it is safe to do so.

When citing previous motor skill acquisition work, Hochmitz and Yuviler-Gavish (2011) indicate "transfer from the first task (the simulation) to the second, real-world task occurs most efficiently when the two tasks have identical component activities. The more elements that are shared between the two tasks, the better the transfer" (p. 490). An integration of physical, functional, and cognitive fidelity during all egress survival training under all possible conditions would ensure that individuals have the opportunity to practice the necessary skills and correct errors in performance under realistic conditions. Regardless of the environmental condition, skill acquisition research indicates that individuals should be required to demonstrate competence through whole-task demonstrations and, if future HUET programs incorporated this type of requirement, it might be possible to better enhance survivability of a real ditching.

REFERENCES

Adams, J. A. (1987). Historical review and appraisal of research on the learning, retention, and transfer of human motor skills. *Psychological Bulletin, 101*(1), 41–74.

Allen, J. A., Hays, R. T., & Buffordi, L. C. (1986). Maintenance training, simulator fidelity, and individual differences in transfer of training. *Human Factors, 28*, 497–509.

Anderson, G. (2008). *CPR and first aid skill retention.* Work Safe BC – RS2006-IG06.

Baldwin, T. T., & Ford, J. K. (1988). Transfer of training: a review and direction for future research. *Personal Psychology, 41*(1), 63–105.

Barreiros, J., Figueiredo, T., & Godinho, M. (2007). The contextual interference effect in applied settings. *European Physical Education Review, 13*(2), 195–208.

Battig, W. F. (1979). The flexibility of human memory. In L. S. Cernak, & F. J. M. Craik (Eds.), *Levels of processing and human memory*. Hillsdale, NJ: Erlbaum. pp. 23–44.

Bell, B., Szczepkowski, M., Santarelli, T., & Schlachter, J. (2005). Cognitive fidelity as a touchstone for on-demand team training. In *Behavior representation in modeling and simulation conference proceedings* (05-BRIMS-052).

Bradley, P. (2006). The history of simulation in medical education and possible future directions. *Medical Education, 40*, 254–262.

Cheung, B., Hofer, K., Brooks, C., & Gibbs, P. (2000). Underwater disorientation as induced by two helicopter ditching devices. *Aviation, Space, and Environmental Medicine, 71*(9), 879–888.

Civil Aviation Authority. (1995). *CAP 641 – Report of the review of helicopter offshore safety and survival.* Report available from: https://www.caa.co.uk/docs/33/CAP641.PDF.

Civil Aviation Authority. (2014). *Safety review of offshore public transport helicopter operations in support of the exploitation of oil and gas.* Report CAP 1145.

Coleshaw, S. R. K. (2006). *Investigation of removable exits and windows for helicopter simulators.* OPITO Report SC 153.

Coleshaw, S. R. K. (2013). *Goals for basic survival training. CNLOPB report.* Report number SC 181.

Cooke, N. J., Gorman, J. C., Myers, C. W., & Duran, J. L. (2013). Interactive team cognition. *Cognitive Science, 37*(2), 255–285.

Cunningham, W. F. (1978). Helicopter underwater escape trainer (9D5). In *AGARD conference proceedings, No. 255 (Operational helicopter aviation medicine)*, 66-1–;66-3.

Gerathewohl, S. J. (1969). Fidelity of simulation and transfer of training: a review of the problem. In *Paper presented at AGARD-NATO – Advanced operational aviation medicine.*

Hays, R. T. (1981). *Training simulation fidelity guidance: The iterative data base approach. U.S. Army Research Institute for the Behavioral and Social Sciences.* Technical report 545.

Hays, R. T., Jacobs, J. W., Prince, C., & Salas, E. (1992). Flight simulator training effectiveness: a meta-analysis. *Military Psychology, 4*, 63–74.

Hays, R. T., & Singer, M. J. (1989). *Simulation fidelity in training system design: Bridging the gap between reality and training.* New York: Springer-Verlag.

Hochmitz, I., & Yuviler-Gavish, N. (2011). Physical fidelity versus cognitive fidelity training in procedural skills acquisition. *Human Factors, 53*(5), 489–501.

Hossein, H., Fallah, A., & Raza, M. (2012). New role for astroglia in learning: formation of muscle memory. *Medical Hypotheses, 79*, 770–773.

Joiner, W. M., & Smith, M. A. (2008). Long-term retention explained by a model of short-term learning in the adaptive control of reaching. *Journal of Neurophysiology, 100*, 2948–2955.

Kaiser, M. K., & Schroeder, J. A. (2003). Flights of fancy: the art and science of flight simulation. In M. A. Vidulich, & P. S. Tsang (Eds.), *Principles and practice of aviation psychology*. Mahwah, NJ, US: Lawrence Erlbaum Associates, Publishers. pp. 435–471.

Khan, M. A., & Franks, I. M. (2004). The utilization of visual feedback in the acquisition of motor skills. In A. M. Williams, & N. J. Hodges (Eds.), *Skill acquisition in sport: Research, theory, and practice*. New York: Routledge.

Knass, K. (2013). *How I survived North sea helicopter crash that killed four – Hull man Paul Sharp*. Retrieved from: http://www.hulldailymail.co.uk/Holderness-Road-man-escaped-death-helicopter/story-20093886-detail/story.html#ixzz2qaRpl0Qo.

Kozey, J. (2007). *Man rating the helicopter underwater escape trainer using an advanced survival systems jib crane at three different descent speeds. Survival systems limited technical report.*

Kozey, J., McCabe, J., & Jenkins, J. (2007). The effects of different training methods on egress performance from the modular egress training simulator. In *Safe conference proceedings*.

Krakauer, J. W., Ghilardi, M.-F., & Ghez, C. (1999). Independent learning of internal models for kinematic and dynamic control. *Neuroscience Nature, 2*(11), 1026–1031.

Kreibig, S. D. (2010). Autonomic nervous system activity in emotion: a review. *Biological Psychology, 84*, 394–421.

Lazurus, R. S., & Folkman, S. (1984). *Stress, appraisal, and coping*. New York: Springer. pp. 117–140.

Lee, T. D., & Simon, D. A. (2004). Contextual interference. In A. M. Williams, & N. J. Hodges (Eds.), *Skill acquisition in sport: Research, theory and practice*. New York: Taylor and Francis. pp. 29–44.

Lim, J., Reiser, R. A., & Olina, Z. (2009). The effects of part-task and whole-task instructional approaches on acquisition and transfer of complex cognitive skill. *Educational Technology Research and Development, 57*, 61–77.

Meichenbaum, D., & Deffenbacher, J. L. (1988). Stress inoculation training. *The Counseling Psychologist, 16*(1), 69–90.

Mills, A. M., & Muir, H. (1999). Development of a standard for underwater survival. In *Technical paper for Shell Group*.

Muscolino, J. (2006). What is muscle memory? the posture of our body is largely determined by the memorized pattern of baseline tone of the musculature of our body. *Message Therapy Journal, 48*(1), 105–108.

NUTEC-MTC. (2001). *Safety alert NUTEC-MTC group*. Report No. 001/SHEQ.

Panzer, S., Wilde, H., & Shea, C. H. (2006). Learning of similar complex movement sequences: proactive and retroactive effects on learning. *Journal of Motor Behavior, 38*, 60–70.

Pauwels, L., Swinnen, S. P., & Beets, I. A. M. (2014). Contextual interference in complex bimanual skill learning leads to better skill persistence. *PLoS ONE, 9*(6), e100906.

Pine, Z. M., Krakauer, J., Gordon, J., & Ghez, C. (1996). Learning of scaling factors and reference axes for reaching movements. *Neuroreport, 7*, 2357–2361.

Porter, J. M., & Magill, R. A. (2010). Systematically increasing contextual interference is beneficial for learning sport skills. *Journal of Sports Science, 28*(12), 1277–1285.

Robertson, E. M., Tormos, J. M., Maeda, F., & Pascual-Leone, A. (2001). The role of the dorsolateral prefrontal cortex during sequence learning is specific for spatial information. *Cerebral Cortex, 11*(7), 628–635.

Robinson, S. J., Sunram-Lea, S., Leach, J., & Owen-Lynch, P. J. (2008). The effects of exposure to an acute naturalistic stressor on working memory, state anxiety and salivary cortisol concentration. *International Journal on the Biology of Stress, 11*(2), 115–124.

Rose, D., & Christina, R. (2006). *A multilevel approach to the study of motor control and learning* (2nd ed.). San Francisco, CA: Pearson/Benjamin Cummings.

Rose, F. D., Attree, E. A., Brooks, B. M., Parslow, P. R., Penn, P. R., & Ambihaipahan, N. (2000). Training in virtual environments: transfer to real world tasks and equivalence to real tasks. *Ergonomics, 43*(4), 494–511.

Ryack, B. L., Luria, S. M., & Smith, P. F. (1986). Surviving helicopter crashes at sea: a review of studies of underwater egress from helicopters. *Aviation, Space, and Environmental Medicine, 57*(6), 603–609.

Schmidt, R. A. (1975). A schema theory of discrete motor skill learning. *Psychological Review, 82*(4), 225–260.

Schmidt, R. A. (2003). Motor schema theory after 27 years: reflections and implications for a new theory. *Research Quarterly for Exercise and Sport, 74*(4), 366–375.

Schmidt, R. A., & Bjork, R. A. (1992). New conceptualizations of practice: common principles in three paradigms suggest new concepts for training. *Psychology Science, 3*(4), 207–217.

Schmidt, R. A., & Lee, T. D. (2011). *Motor control and learning.* Champaign, IL: Human Kinetics.

Schmidt, R. A., & Wrisberg, C. A. (2008). *Motor learning and performance: A situation-based learning approach.* Champaign, IL: Human Kinetics.

Shea, J. B., & Morgan, R. L. (1979). Contextual interference effects on the acquisition, retention, and transfer of a motor skill. *Journal of Experimental Psychology: Human Learning and Memory, 5*, 179–187.

Shusterman, R. (2011). Muscle memory and the somaesthetic pathologies of everyday life. *Human Movement, 12*(1), 4–15.

Simmons, R. R., Lees, M. A., & Kimball, K. A. (1978). *Visual performance/workload of helicopter pilots during instrument flight.* U.S. Army Aeromedical Research Laboratory. USAARL Report No. 78-6.

Stokes, P. D., Lai, B., Holtz, D., Rigsbee, E., & Cherrick, D. (2008). Effects of practice on variability, effects of variability on transfer. *Journal of Experimental Psychology, 34*(3), 640–659.

Summers, F. (1996). *Procedural skill decay and optimal retraining periods for helicopter underwater escape training. IFAP technical report.*

Taber, M. J. (2010). Offshore helicopter safety inquiry report. In R. W. Wells (Ed.), *(Commissioner), Canada-Newfoundland and Labrador offshore helicopter safety inquiry* (Vol II). pp. 211–290.

Taber, M. J. (2013). Crash attenuating seats: effects on helicopter underwater escape performance. *Safety Science, 57*, 179–186.

Taber, M. J. (2014a). Simulation fidelity and contextual interference in helicopter underwater egress training: an analysis of training and retention of egress skills. *Safety Science, 62*, 271–278.

Taber, M. J. (2014b). *Post-egress survival skills in low level light conditions. Technical report for Petroleum Research Newfoundland and Labrador.*

Taber, M. J., Kozey, J., & McCabe, J. (2012). *Investigation of emergency breathing apparatus skill set knowledge transfer between helicopter underwater escape training simulators.* Unpublished technical document for survival systems training limited.

Taber, M. J., & McGarr, G. W. (2013). Confidence in future helicopter underwater egress performance: an examination of training standards. *Safety Science, 60*, 169–175.

Taber, M. J., & Sweeney, D. H. (2014). Forces required to jettison a simulated S92 passenger exit: optimal helicopter underwater egress training techniques. *International Journal of Industrial Ergonomics, 44*, 544–550.

Thomas, M. J. W. (2003). Operational fidelity in simulation-based training: the use of data from threat and error management analysis in instructional systems design. In *Proceedings of SimTecT 2003: Simulation conference*. pp. 91–95.

Thorndike, E. L., & Woodworth, R. S. (1901). The influence of improvement in one mental function upon the efficiency of other functions. *Psychological Review, 8*, 247–261.

Tipton, M. J., Gibbs, P., Brooks, C., Roiz de Sa, D., & Reilly, T. (2010). ECG during helicopter underwater escape training. *Aviation, Space & Environmental Medicine, 81*, 399–404.

Transportation Safety Board of Canada. (2009). *Determination of impact acceleration and orientation: Helicopter, Sikorsky S-92A, C-GZCH.* Engineering Report LP092/2009.

Transportation Safety Board of Canada. (2010). *Main gearbox malfunction/collision, with water.* Aviation Investigation Report A09A0016. Ministry of Public Works, and Government Services.

Wells, R. (2010). *Canada-Newfoundland and Labrador offshore helicopter safety inquiry* (Vol. 1). St. John's, NL: Canada-Newfoundland and Labrador Offshore Petroleum Board.

Wohldmann, E. L., Healy, A. F., & Bourne, L. E. (2008). A mental practice superiority effect: Less retroactive interference and more transfer than physical practice. *Journal of Experimental Psychology: Learning, Memory, and Cognition, 34*(4), 823–833.

Woolfson, C., Foster, J., & Beck, M. (1996). *Paying for the piper: Capital and labour in Britain's offshore oil industry.* London: Mansell Publishing Limited.

Wright, D., Verwey, W., Buchanen, J., Chen, J., Rhee, J., & Immink, M. (2015). Consolidating behavioral and neurophysiologics findings to explain the influence of contextual interference during motor sequence learning. *Psychonomic Bulletin & Review, 22*(3), 1–21.

Yerkes, R. M., & Dodson, J. D. (1908). The relation of strength of stimulus to rapidity, of habit-formation. *Journal of Comparative Neurology and Psychology, 18*, 459–482.

CHAPTER 8

Helicopter Emergency Breathing Systems

Sue Coleshaw
Independent Research Consultant, Kincardine O'Neil, Aboyne, Aberdeenshire, UK

8.1 INTRODUCTION—THE NEED FOR EMERGENCY BREATHING SYSTEMS

Accident statistics over the years have shown that drowning is the main cause of death in helicopter water impact accidents (Chen, Muller, & Fogarty, 1993; Clifford, 1996; Rice & Greear, 1973). This is perhaps not surprising given the fact that helicopters capsize or sink in a high proportion of cases, either immediately or shortly after impact, meaning that those who survive the impact must then escape through exits that are now under water. At the same time, the individual must cope with inrushing water and the disorientation caused by inversion (Brooks, 1989; Cheung, Hofer, Brooks, & Gibbs, 2000; Jamieson, Armstrong, & Coleshaw, 2001; Ryack, Luria, & Smith, 1986). Without any personal equipment to help, survival depends on the person being able to take a breath before the head goes under water and then completing an underwater escape within the breath-hold time of that individual. It is generally accepted that in cold water there is a mismatch between breath-hold time and the time that may be required to effect an escape from the helicopter (Cheung, D'Eon, & Brooks, 2001; Coleshaw, 2003, 2013; Tipton, Balmi, Bramham, Maddern, & Elliot, 1995). This is particularly true if the individual must cross the cabin to reach a usable exit or wait for another person to escape before being able to use the nearest exit. Equipment designed to extend the time that can be spent under water will increase the likelihood of escape and, it is hoped, reduce the incidence of drowning.

The prevalence of drowning differs depending upon the type of water impact. Few lives have been lost in a controlled ditching (defined as an emergency landing on water deliberately executed by the crew). Under these circumstances Clifford (1996) reports only four cases of drowning,

Handbook of Offshore Helicopter Transport Safety
ISBN 978-1-78242-187-0
http://dx.doi.org/10.1016/B978-1-78242-187-0.00008-0

where cause of death was known, out of a total of 308 occupants (cited in Coleshaw, 2003). In most cases of controlled ditching the aircraft remains afloat and upright for sufficient time to safely evacuate the helicopter occupants into a life raft (e.g., AAIB, 1997, 2014; AAIB/N, 1998). Capsize is still probable in higher sea states, but it is less likely to occur immediately. Further, the crew may be less inclined to execute an emergency landing in high sea states. A post-ditching capsize is more likely when external flotation is not provided, is not deployed, or is insufficient for the wave conditions, in which case the risk of drowning will be increased.

While impact injuries predominate as the primary cause of fatality in uncontrolled water impact accidents, drowning is the main cause of fatality in potentially survivable water impacts due to a controlled flight into water or vertical descent with limited control (Clifford, 1996; cited in Coleshaw, 2003). In such cases, the occupants may have little or no warning and therefore little time to prepare for the event (Brooks, MacDonald, Donati, & Taber, 2008; Chen et al., 1993). The report following the accident near the Cormorant Alpha platform in the North Sea comments on the ability of the occupants to take a breath of air before head submersion: "water ingress to the cabin was rapid and, although the survivors who had been seated to the rear reported that they had time to take a deep breath of air, those at the front did not" (AAIB, 1993, p. 28). Four individuals survived the impact and released their seat belts but failed to successfully escape to the surface. The aircraft was thought to have capsized but remained at the sea surface for 1–2 min, fully sinking within 4 min of the impact. Breath-holding time was thought to have been the limiting factor that prevented them from making an escape (in water at 7–8 °C). Interestingly, an industry review of safety following this accident (CAA (Civil Aviation Authority), 1995) concluded that there was no clear advantage to be gained from the introduction of an emergency breathing system (EBS) and that the chances of successful underwater escape might be more reliably improved by measures aimed at facilitating egress, placing emphasis on systematic improvements rather than relying on the individual and his or her personal protective equipment. The latter opinion is no doubt true, but in the absence of systematic improvements, EBS can help to reduce the level of risk.

More recently the investigation into an accident off the east coast of Canada suggested: "it is likely that several of the occupants remained conscious after the impact with the water, eventually breaking their breath hold and drowning before they could egress from the sinking helicopter" (TSB, 2010, p. 128). Drowning was given as the cause of all 17 fatalities.

This is another example of an accident in which EBS could have saved lives. A compressed-air system was introduced shortly after the accident, the culmination of several years' work by the industry to select a product.

A major contributor to the high incidence of drowning is cold shock (see Chapters 4 and 5 for further detail), which significantly reduces the breath-hold time of helicopter occupants, limiting their chances of making a successful escape from the aircraft. The effects of cold on breath-holding have been implicated as the cause of drowning in a number of helicopter accidents (Brooks, 1989; TSB, 2010).

Breath-hold duration following sudden immersion decreases linearly with a reduction in water temperature (Hayward, Hay, Matthews, Overweel, & Radford, 1984). In a relatively benign water temperature of 24 °C the breath-hold time of 228 civilian and military offshore workers ranged from 5 to 120 s, with a median of 37 s (Cheung et al., 2001). At water temperatures less than 15 °C, breath-hold times of individuals not protected by a suit may be less than 30 s and only 25–50% of presubmersion levels (Hayward et al., 1984). Even when wearing a helicopter suit (see Chapter 9 for a detailed discussion of helicopter passenger suits) the evidence shows that immersion in water colder than 10 °C results in maximum breath-hold times of less than 20 s and sometimes as low as 6 s (Tipton et al., 1995; Tipton, Franks, Sage, & Redman, 1997; Tipton & Vincent, 1989).

These progressively shorter breath-hold durations allow very little time for an individual to escape from a capsized helicopter. Research suggests that egress from an underwater egress simulator (UES) with a seat next to an exit takes 25–30 s (Bohemier, Chandler, & Gill, 1990; Coleshaw & Howson, 1999). Helicopter simulator trials conducted with a full complement of passengers have shown escape times ranging from 27 to 92 s (Brooks, Muir, & Gibbs, 1999); the last ones out had to use EBS to complete their escape. These times can be compared with the Cheung study when 34% of subjects had a breath-hold time of less than 28 s (Cheung et al., 2001). The disparity between breath-hold time in cold water and the time that may be required to escape from a helicopter provides a strong justification for providing emergency underwater breathing equipment.

If EBS can be used effectively in a capsize accident, then the extended time that a person can remain under water is beneficial in providing sufficient time to carry out all the actions needed to locate a usable escape route, exit from the helicopter, and reach the water surface. Such actions will have to be undertaken when the individual is feeling disoriented following inversion, is in reduced-light conditions, and is coping with many other factors that could hinder egress. The person might struggle to release the seat harness because of the stress of the situation or possibly uneven loading on the harness. Release of the harness could also be hampered by other personal equipment. Following water impact, the airframe may well be deformed; this might prevent the operation of an exit door, or reduce the size of an exit, meaning that the nearest or most obvious escape route is blocked. Seat backs may be deformed, making escape difficult. An escape route may be blocked by another person who is unconscious, injured, or behaving in an irrational manner. This could take the form of inaction or an inappropriate reaction (Leach, 2004; Muir, 1999—see also Chapter 3). In each of these situations rapid escape is hindered and the person is unlikely to be able to make a successful escape when depending upon a breath-hold. With the aid of EBS, it is hoped that the person will be able to breathe under water for sufficient time to overcome the obstacle, locate a usable exit route, and egress from the helicopter.

The benefits of using EBS were reported by Ryack, Smith, Champlin, and Noddin (1977) following research into the illumination of helicopter escape hatches. Twenty-four test subjects, all experienced divers, were given a prototype compressed-air breathing device containing 4.5 min of air supply in case they became disoriented during the underwater escape. There were 16 occasions on which the subject became disoriented, lost, and/or entangled, all but one in the condition without illumination. In eight of these cases, the subject relied on the breathing device to get him or her out of difficulty. Two further subjects failed to use the device successfully. The authors supported the provision of underwater breathing devices and it was later concluded that the provision of breathing equipment to helicopter occupants would allow crash survivors "adequate time to act appropriately" (Ryack et al., 1986, p. 608).

Taber and McCabe (2009) recommended issue and training in the use of EBS for troops flying over water, after underwater escape trials in which only 58% of subjects successfully escaped on a breath-hold, whereas 100% were able to escape when using EBS. Further, they described how subjects were able to egress through a partially blocked exit when using EBS. Although the subjects took longer to escape, they rated their performance as being improved when using EBS. This demonstrates how EBS can allow the user time to think through actions and follow correct procedures when making an escape.

While there are only a limited number of helicopter accidents in which EBS have been used, those reports from survivors that exist suggest that the use of EBS has a calming effect on the user (Barker, Yacavone, Borowsky, & Williamson, 1992), and feelings of relief were experienced when the EBS was successfully deployed (see Brooks & Tipton, 2001; Coleshaw, 2003), allowing the user to overcome the understandable feelings of panic.

It can therefore be concluded that there is a very clear need for the use of EBS by helicopter occupants flying over water. That said, it is undesirable to rely on personal protective equipment and on the ability of the person to use the equipment correctly under the stresses of an emergency. It would be preferable to achieve systematic improvements that reduce the likelihood of occupants having to make an underwater escape. While such improvements in helicopter safety may reduce the need for EBS in the future, at the present time there is a strong case to say that EBS should be issued to helicopter occupants, particularly those who have to fly over hostile sea areas.

8.2 HISTORICAL DEVELOPMENT

8.2.1 Early Development of EBS for Use by Military and Coast Guard Aircrew

It has long been recognized that lives are lost from drowning when attempting to escape from a submerged aircraft. The concept of EBS was first investigated as early as the 1940s. Some product testing of EBS by the US Navy is believed to have taken place as early as 1945 (cited in Brooks & Tipton, 2001), with the first compressed-gas products thought to have been brought into full service in the early 1980s (Figure 8.1).

In 1963, a report from the US Navy Mine Defence Laboratory (Odum, 1963) described the design and testing of an emergency breathing apparatus fitted into a life jacket. This was a semi-closed system delivering a 75% oxygen gas mixture; open circuit systems were considered to be inefficient and would need too large a tank to provide sufficient gas. A 15-min breathing supply was provided, considerably longer than is necessary for helicopter underwater escape. The report recommended use of the EBS by the crew of US Navy aircraft, including helicopters, as an aid to survival in the event of a crash into water.

The underwater escape rebreather vest, developed for the US Coast Guard in the early 1980s, was also a dual-purpose system with the rebreather incorporated into an inflatable life jacket (Figure 8.2). This device also used oxygen as the breathing gas. An evaluation by the US Navy Experimental Diving Unit concluded that a 40% oxygen mixture provided the best combination of breathing duration and safety (Gray, Thalmann, & Syklawer, 1981).

Figure 8.1 Early version of compressed-air EBS used by the Italian Navy c. 1985. *(Image courtesy of Dr. Chris Brooks with permission.)*

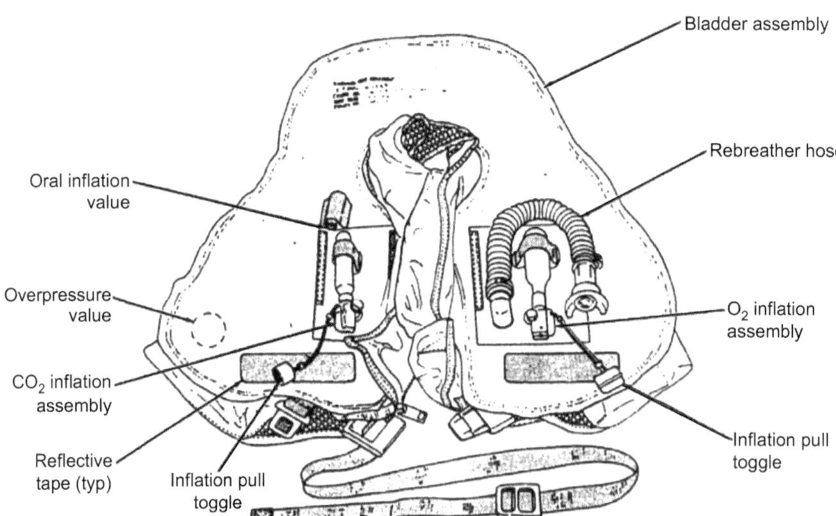

Figure 8.2 US Coast Guard underwater escape rebreather. *(Image from Brooks and Tipton (2001) with permission.)*

However, perhaps because the maximum duration of use of the 40% oxygen device when exercising in cold water was only 2 min, the unit that went into service with the US Coast Guard provided 100% oxygen. The operational system was later criticized for using pure oxygen (Brooks, 1989; Hayes, 1991). There was concern that there would be a lack of stimulus to respiration owing to a lowered hypoxic drive and a tendency for severe hypercapnia to develop. This would lead to loss of consciousness. Hayes reported collapse of the hose over the entry point to the rebreather, with certain positions in the water worse than others. Once brought into operation by the US Coast Guard further problems were reported. The lack of a demand valve meant that the gas supply could be lost if the unit was knocked out of the mouth. The buoyancy of the inflated gas bladder was also an issue; during training it was found that egress was possible when hand-over-hand contact was made but difficulty was experienced if a handhold was lost. It was believed that this was one reason the system was later rejected by the US Navy (Coleshaw, 2003, p. 24). The US Coast Guard were themselves concerned about the fire risk associated with the use of pure oxygen, leading them to later switch to a compressed-air system.

An alternative helicopter EBS incorporating a gas bottle, regulator, mouthpiece, and purge button, using compressed air rather than oxygen, was investigated by the Canadian Defence and Civil Institute of Environmental Medicine and by the US Navy in the early 1980s (Brooks & Tipton, 2001). This device was a relatively simple system based on a design originally developed in 1979 to aid sport divers who had depleted their main gas supply, providing them with sufficient air to reach the surface.

Following extensive trials, the US Navy brought the so-called HEED 2 (helicopter emergency egress device) into service in 1986 (Figure 8.3). The benefit of this compressed-air EBS was proven as early as 1987 when a helicopter crashed into the Pacific and two aircrew members used the compressed-air EBS to make a successful escape (Brooks & Tipton, 2001). The Canadians meantime made modifications to the device with the addition of a hose between the cylinder and the regulator, allowing the gas cylinder to be stored in their survival backpack. After a few further modifications, this device was brought into service in Canada in 1988. Following some servicing problems and as new systems came onto the market, a further procurement process was undertaken in 1993–1994 with five compressed-air EBS considered, some with and some without a hose (Brooks & Tipton, 2001). After extensive testing, the product chosen could be operated with one hand and incorporated a hose so that it could be mounted over the shoulder on the back of the life vest.

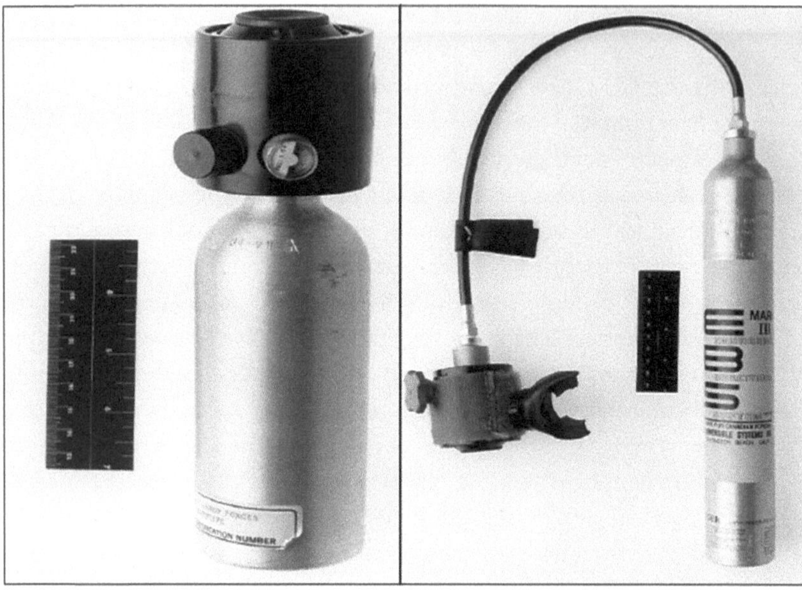

Figure 8.3 HEED 2 and modified Canadian version with hose. *(Image from Brooks and Tipton (2001) with permission.)*

The first EBS considered by the UK Royal Navy was the HEBE (helicopter emergency breathing equipment), which was trialed in 1975 (Brooks & Tipton, 2001), but is not thought to have gone into service. It was not until 1989 that a procurement process was put in place for a "short-term air supply system" (STASS), following a Royal Air Force report (Sowood, 1989) which demonstrated that an air supply significantly aided in escape from a submerged helicopter. STASS was required to assist trained military aircrew in their escape from the helicopter in the event of a forced landing on water. The initial specification included the requirements for a compressed-air system to be mounted on the life jacket, have first-breath activation, have a capacity of $\geq$50 L at 207 bar, have the capability to be purged in any orientation, have a contents gauge, and be able to withstand the forces of ditching and remain fully functional (Coleshaw, 2003). In 1992 training was commenced and the STASS was introduced into service for the Royal Navy aircrew and for Royal Air Force crew flying over water.

A year later, work started to identify equipment suitable for military passengers (see Coleshaw, 2003). Owing to the size, nature, and geographic distribution of the potential passenger population, it was considered that in-water training would be impractical. The aim was to provide an EBS that would be very easy to use and intuitive, such that the training of passengers could be limited to a briefing prior to making a flight. The product selected, a compressed-air device carried in a pouch around the waist, had a specially modified mouthpiece that would limit water entry and thus reduce the need for purging.

8.2.2 Development of EBS for Use by Offshore Passengers

The first EBS to be introduced for civilian helicopter passengers was a rebreather system developed for the UK offshore workforce traveling to and from oil and gas installations in the North Sea. The rebreather system was initially brought into service by two operators in 1996. It was integrated into the passenger suit system, but when this was found to restrict the counterlung it was modified to be carried in a pouch that sat on the chest between the lobes of the passenger life jacket. In 1999 the product was further modified and replaced by a hybrid rebreather system. This consisted of the original rebreather system, but with a small (3 L) cylinder of air that would be automatically discharged into the counterlung on immersion. This meant that a user who did not manage to take a breath before submersion would still have air in the bag to breathe when under water. It was only when this second-generation hybrid product was launched that EBS was adopted across the UK offshore industry.

Since the introduction of the hybrid rebreather in the United Kingdom, further products have been introduced for civilian offshore workforce passengers in other jurisdictions. Norway introduced a rebreather integrated into their passenger suit system, meeting the requirements of an Oljearbeidernes Fellessammenslutning (Norwegian Oil Industry Association) specification (OLF, 2004). This specification required the breathing system to be automatically activated on submersion, reducing the actions that must be remembered and undertaken by the user.

The Canadian offshore industry was the first to introduce a compressed-air design for civilian passengers in 2009. The product selected was similar to designs used by the military (compressed-air EBS being the product of choice for military operations). Introduction for use by passengers had been delayed owing to concerns over a small risk of lung barotrauma during training. Prior to a fatal accident in 2009 (TSB, 2010), a program to select an EBS product for the Canadian offshore industry had been initiated but the equipment was not introduced until after the accident. Training was successfully brought in (Brooks, MacDonald, Carroll, & Gibbs, 2010) but was limited to exercises in shallow water with the aim of reducing any risk of barotrauma to a negligible level.

The carriage of EBS by passengers working offshore was until 2015 mandated and managed by the oil and gas operators who employ the workforce. There had been little or no regulation of EBS usage and no regulatory requirement for civilian passengers or crew to be equipped with EBS for overwater flights. This changed when, in 2013, the Government of Canada published proposals to amend the Canadian Aviation Regulations, mandating the use of an emergency underwater breathing apparatus for each person on board an offshore operations flight over Canadian waters. This amendment was published in April 2015 (Government of Canada, 2015).

Following a number of helicopter accidents in the UK sector between 2009 and 2013, serious concerns led to a review of safety being initiated by the UK Civil Aviation Authority in September 2013 (CAP 1145, CAA, 2014a). One of the many and various conclusions was the need to mandate EBS. Their required action stipulated that passengers on UK offshore flights must wear an EBS that met Category "A" of the technical standard detailed in CAP 1034 (Coleshaw, 2013). The only circumstances in which this would not be required would be if the helicopter occupant was sitting in a seat next to an exit or the helicopter was fitted with flotation equipment that would prevent complete capsize. The overall aim of this action was to give occupants a better chance of escape. The action was regulated by an Airworthiness Directive in October 2014 (CAA, 2014b) and was brought into force for passengers in January 2015. The rule will apply to all occupants including helicopter crew from April 2016. The regulatory action resulted in two significant outcomes. The first was for the UK offshore industry to switch to using a compressed-air EBS rather than the hybrid rebreather previously carried, the new product allowing a faster deployment time and having a greater capability for underwater deployment following capsize. The second was a proposal for EBS use by helicopter occupants in offshore operations to be mandated by the European Aviation Safety Agency (EASA). Amendments to the European Rules relating to helicopter offshore operations have been proposed (EASA, 2014). This could have greater implications worldwide if global operators apply the new rules across their operations.

8.3 DESIGN

8.3.1 General

The varying development of emergency underwater breathing systems for helicopter occupants has meant that EBS fall into two main design types: compressed-gas systems that provide breathable gas on demand and systems that rely on rebreathing the air in the lungs. Hybrid designs that consist of a rebreather system with additional compressed gas also exist. Each design type has differing advantages and disadvantages which must be evaluated by the end user when selecting a device that will be suitable for particular conditions of use.

8.3.2 Compressed-Air EBS

Most of the compressed-gas systems now on the market use air as the breathable gas. They can be used at depths of 10 m or more and have been designed to be capable of deployment under water. They do not rely on the user having time to take a deep breath of air before submersion to extend the time spent under water (a key disadvantage of rebreather systems). This means that they are particularly suitable for use in accidents when capsize or sinking occurs immediately after contact with the water. (It must be remembered that if deploying under water, the user still has to breath-hold for the time it might take to deploy the device, possibly after overcoming the disorientation caused by a capsize.)

Breathing times of 3–5 min (Brooks & Tipton, 2001) may be achieved, with duration of use dependent upon the size of the cylinder, conditions of use, and ability of the user to control his or her breathing. In the worst case, a gas bottle may be breathed down in less than a minute if, for example, used in very cold water by an individual who is highly stressed and breathing heavily. Because of this limited gas supply, compressed-gas systems should not be deployed too soon in anticipation of submersion. In warmer water and with breathing under control the same gas supply will last for a much longer period. One further factor to consider is that the gas can run out with little or no warning, so the user must know how to deal with this situation.

A disadvantage of compressed-gas systems is the small risk of lung barotrauma (injury caused by rapid or excessive pressure changes) that can occur during rapid ascent. Barotrauma can be caused by expanding air in the lungs that is not exhaled during ascent. The level of risk in shallow water is poorly defined. There are a few isolated case reports of problems caused by lung overpressure during military helicopter underwater escape training (HUET) using compressed-air EBS (Benton, Woodfine, & Westwood, 1996; Risberg, 1997). The first was a case of arterial gas embolism, the second a case of mediastinal emphysema; in both cases the individuals made a full recovery. While the use of compressed air does present a slightly increased risk of lung barotrauma, it should be noted that some similar lung injuries have also been observed during breath-hold dives, possibly due to localized obstruction of the airways or coughing when under water (Henckes, Arvieux, Cochard, Jézéquel, & Arvieux, 2011; Shah, Thomas, & Gibb, 2007). Less serious ear problems have also been reported (Risberg, 1997; cited by Coleshaw, 2003). The effects of pressure change on the ear or sinuses are similar whether using compressed-air or rebreather EBS. Barotrauma is a potential problem only for training as the risk is accepted as being negligible in relation to a real accident.

While most designs of compressed-gas EBS are similar in that they consist of a gas cylinder, first-stage regulator, demand valve (second-stage regulator), and mouthpiece, some have the first- and second-stage regulators very close together, whereas others have a medium pressure hose between the regulators. The products with a hose have the advantage that the relatively heavy gas cylinder can be held securely in a convenient position that will not hinder movement and only the weight of the mouthpiece and hose has to be held by the teeth. If the grip on the mouthpiece is lost the device is still held securely and the user only has to locate the cylinder and hose to find the mouthpiece and redeploy. If a device without a hose is knocked from the mouth there is a danger that the whole unit will be lost unless attached to the body in some way.

Overall, while care must be taken when considering the training requirements for compressed-air EBS, their capability to be quickly deployed under water has meant that they are favored by many user groups and they are known to have saved lives in a number of military helicopter accidents (see Brooks & Tipton, 2001 for examples).

8.3.3 Rebreather EBS

The next EBS design type is the rebreather system, which allows the user to rebreathe air from his or her lungs, exhaling into and inhaling from a counterlung. With a simple rebreather system (rather than a hybrid rebreather, described later), the user must have time to take a deep breath prior to head immersion for the device to perform to its full capability. If there is no time to take a deep breath the user will be dependent upon the volume of air already in the lungs, the worst case being no air to exhale into the counterlung. It goes without saying that in this scenario a rebreather will provide no benefit to the user. Simple rebreather systems may therefore have limited capability in accidents in which capsize or submersion occur immediately after impact with the water.

The period of use of a rebreather EBS is limited by the time it takes for the carbon dioxide concentration to build up in the counterlung and for the oxygen concentration to gradually decrease. This will be influenced by the volume of air in the lungs when use of the rebreather is initiated; with a reasonable initial breath in it is possible to rebreathe from a counterlung for several minutes. Another disadvantage of rebreathers is that the user must overcome hydrostatic pressure when exhaling into the counterlung; the deeper the water the more difficult this will be. Their capability is therefore limited at depths greater than 3–5 m. Hydrostatic imbalance, the difference in pressure between the lungs and the counterlung, depends upon the orientation of the user and the position of the counterlung. With a counterlung on the chest, the greatest breathing resistance will be experienced when the user is in the face-down position. As long as breathing resistance is kept within certain limits (see Coleshaw, 2013), it may cause some discomfort but can be tolerated for the period of time required to escape from the helicopter.

Rebreather systems are generally designed so that the user initially breathes the atmosphere and then switches to breathing into and out of the counterlung. This means that, when time allows, the rebreather can be deployed early, with the switch to breathing from the counterlung made immediately before immersion. This system has the disadvantage that there is an additional action that will add several seconds to deployment time when early deployment is not feasible. Some systems make the switch automatic (by water activation). This removes the additional action on the part of the user but could allow water entry into the counterlung if the user is submerged before the mouthpiece has been deployed.

8.3.4 Hybrid EBS

Hybrid rebreather devices have the basic components of a rebreather system but with the addition of a small cylinder of compressed gas (air) that can be automatically discharged into the counterlung on immersion. Hybrid systems have the advantage over a simple rebreather that they do not rely on the user taking a deep breath of air before submersion; the discharged air is equivalent to a normal lung-full of air. The system will therefore still be functional even if the user does not have time to take a breath. This will be a big advantage in accidents in which there is little warning and rapid capsize. The additional air will also extend the duration of use. One disadvantage of the additional air provided in a hybrid system is that the overall buoyancy of the user will increase compared to other EBS categories. This can make escape a little more difficult, although the user may have sufficient time to overcome this buoyancy effect if the EBS is functioning correctly (see Chapter 9 for a discussion of escape buoyancy).

When used with the additional cylinder of compressed air, a hybrid EBS will present a small risk of lung barotrauma similar to the compressed-air EBS designs. This can be avoided in training by using the EBS as a simple rebreather without the gas cylinder. If this is done, the users must understand the difference between the training and the operational devices.

8.4 EBS PERFORMANCE

8.4.1 General

When selecting an EBS it is important to know that the equipment will perform effectively in the operating environment, that it is easy to use, and that there are no inherent safety issues that might add to the level of risk.

8.4.2 Deployment

A simple and intuitive design is probably the most important aspect of any piece of emergency-use equipment and this is certainly true for EBS. As previously mentioned, the equipment will be used in a highly stressful situation and often under extreme environmental conditions. In any accident in which there is no warning and capsize occurs almost immediately, there will be little time to prepare and deploy the EBS before submersion. In the event of underwater deployment, this must be achieved within the breath-hold time of the individual and, as has previously been stated, breath-hold times may be less than 10 s in cold water (Tipton et al., 1995; Tipton & Vincent, 1989). Despite this, many specifications for EBS have in the past missed the need for rapid deployment. Likewise, training programs often focus on learning to breathe from the device under water but neglect the practice of rapid emergency deployments.

Factors that can slow down deployment time include pouches that are difficult to open or that hold the EBS too tightly, security ties that cannot be broken, covers that interfere with the easy location of the EBS mouthpiece, and nose clips that are awkward to grasp and fit. There needs to be a single action to remove the mouthpiece from its stowed position and place it in the mouth. Successful operation of the EBS can fail if the mouthpiece is lost during deployment; in this event it must be easy to relocate. There is some debate regarding the value of nose clips as they take time to deploy and do not always fit well, but some users cannot use EBS without the aid of a nose clip, particularly when inverted (Coleshaw, 2013). It is therefore desirable to provide a nose clip or other nose occlusion system with the EBS for those that need it, but deploy this only if considered necessary. Those who do not need a nose clip should not waste time deploying one. A half-face mask could provide nose occlusion and has the added benefit of improving vision under water. (As with a nose clip, there would need to be sufficient time to deploy a half-face mask, and in this case, deployment would preferably take place on the water surface.) The user must be familiar with the EBS and have undertaken training in water to ascertain the need for nose occlusion.

Underwater deployment provides much greater challenges to the user than dry deployment. While most designs will aim to limit water entry into the mouthpiece, some more successfully than others, there is always likely to be some water that must be cleared. Compressed-gas designs provide a purge button that allows some of the gas supply to be used to blow the water out of the mouthpiece. Alternatively, the user can use expired air to blow the water out. The latter method can be used with a rebreather, the only other option being to swallow small amounts of water, although the success of this will depend upon the dead space of the mouthpiece and the volume of water involved. Underwater deployment must be achieved when the individual is coping with the effects of the cold shock response and most likely with the disorientation caused by capsize of the helicopter. Only simple designs are likely to perform well under such circumstances.

8.4.3 Duration of Use

Once deployed, it is necessary to ensure that the EBS will significantly extend the time that can be spent under water. Many user groups specify a minimum duration of use. In doing this, it is advisable to also specify the likely conditions of use, such as water temperature and water depth, as these will influence how long a user can comfortably breathe from the device. Testing in cold water at a temperature appropriate for the operating environment is important as cold shock will make underwater deployment more difficult to achieve. Colder water temperatures will also result in shorter durations of use, as cold shock increases the rate and depth of breathing. Gas supplies will be used up more quickly, while the work of breathing will be increased for both compressed-air and rebreather systems at high levels of ventilation.

8.4.4 Breathing Performance

To ensure that an EBS can be used safely, it is necessary to measure the breathing performance of the equipment. It is known that cold shock results in a high rate and depth of ventilation for the first few minutes of cold-water immersion (Tipton, Stubbs, & Elliot, 1991). Anxiety can also alter breathing patterns, again increasing ventilation through an emotional response (Masaoka & Homma, 1997, 2001), although conversely, use of an EBS has been shown to have a calming effect (Barker et al., 1992) and may help to limit panic breathing. It is, however, important to ensure that the EBS will perform satisfactorily at these high rates of ventilation. A mechanical breathing simulator can be used to measure the work of breathing and breathing resistance, thereby ensuring that the equipment will perform within accepted safe limits. Rebreather systems are likely to produce higher levels of breathing resistance than compressed-air systems. The breathing simulator also allows the work of breathing to be assessed at low water temperatures and reduces the need to expose human test subjects to very cold water (e.g., less than 10 °C).

8.4.5 Compatibility and Integration

It is important that EBS are compatible with and do not impair the performance of aircraft equipment or other personal protective equipment (see Chapter 9 for a detailed discussion of survival suit system integration). The pilots must be able to perform their normal operational activities as well as emergency functions without hindrance or discomfort. The EBS must not interfere with the operation or release of the seat harness, and no part of the EBS should snag during egress through small emergency exit windows. If the EBS adds buoyancy, then it is necessary to ensure that the user is still able to undertake an underwater escape. The buoyancy of suit systems is usually assessed during approval; the added buoyancy due to an EBS needs to be included in this process where appropriate.

Speed of deployment is often dependent upon how well the EBS is integrated into the overall personal survival equipment system based on a helicopter immersion suit and/or life jacket. It must be possible to easily locate the EBS, first time, and without error. Most EBS, or parts thereof, are stored in a pocket on the suit or life jacket. If too tightly held in the stowed position, vital seconds could be lost in releasing the device.

In any personal protective equipment system, one item should never impair the performance of another. Good integration is therefore essential.

8.4.6 Technical Standards to Assess Performance

Until recently, there has been no technical standard for the approval of EBS products allowing levels of performance and safety to be assessed. As of this writing, different products have been developed based on the test programs of the manufacturers, in many cases working with potential end users. This can make it difficult to select or compare products as no minimum performance standards have been established. The breathing performance of some devices has been assessed against diving equipment standards such as EN 250 (CEN, 2014), but neither this standard nor those for rebreathing diving apparatus (e.g., EN 14143) are generally relevant to helicopter EBS.

In 2003, Coleshaw drafted an example technical standard on behalf of the UK Civil Aviation Authority (Coleshaw, 2003). This was later developed into a full technical standard within CAP 1034 (Coleshaw, 2013), following extensive trials to provide performance data relating to the various generic designs of EBS (Barwood, Corbett, Coleshaw, Long, & Tipton, 2010; Coleshaw, 2012, 2013). The standard contained within CAP 1034 covers essential requirements, including speed of deployment and ease of use when maneuvering under water, when inverted, and when undertaking underwater escape. Cold-water performance is measured with human subjects in water at a temperature of 12 °C (with ethical approval) and during breathing performance tests with the product immersed in water at 4 °C.

Two categories of EBS are defined in the standard: Category "A" devices have the capability to be deployed both in air and under water, while Category "B" devices have the capability to be deployed in air. Deployment of Category "A" devices must be possible within the cold-water breath-hold time of users: 10 s is allowed for deployment of the mouthpiece, while full deployment of the EBS must be achieved in 12 s, allowing a little extra time to deploy a nose clip or other nose occlusion system. A slightly longer deployment time, 20 s, is considered acceptable for Category "B" devices.

This standard has been submitted to the EASA for possible publication as a European Technical Standard Order.

8.5 TRAINING

8.5.1 Level of Training: Dry or in Water?

How much training to provide in the use of emergency equipment is a problematical issue, particularly if there is perceived to be some risk associated with the use of the equipment under realistic conditions. This is true of EBS, for which the benefits of use are obvious and proven but there is sometimes resistance to the provision of high-fidelity training, perhaps due to concerns about the safety of more difficult parts of the training program.

It can be argued that "dry" training, in which the only practical experience is to deploy and breathe from the EBS, is insufficient training to fully prepare the person for use of the EBS in a real emergency. Some designs of EBS have been developed with the aim of being so intuitive that the minimum training requirement suggested by the manufacturer is a briefing, possibly with a demonstration of its use. This might be successful in showing individuals how to deploy the EBS and prevent significant aspiration of water on immersion, but they will not learn how to achieve a good seal around the mouthpiece or gain skills in how to use the equipment under water when completing an underwater escape sequence.

The first challenge when learning how to use any EBS in water is to overcome the natural instinct, which is to hold one's breath. The body tells us not to breathe and there is therefore a need to overcome this instinctive response. If using a rebreather, the next challenge is to learn to exhale against water resistance. Even with compressed-gas EBS, by which air is supplied on demand, when first used, the initial few breaths constitute an unusual experience and the user must become accustomed to breathing under water. Confidence can be gained after just a few minutes under water, meaning that this initial step in "wet" training is highly valuable. Such experience will also show up problems such as poorly fitting nose clips, problems when achieving a seal around the mouthpiece, and, in a few individuals, an uncomfortable gagging reflex.

While the basic skill of using the EBS under water can be achieved with the head just submerged, this does not provide the user with any experience of the challenges that could be presented in a real emergency environment. These include the ability to use the EBS while maneuvering under water (it is easy to lose grip on the mouthpiece when turning the body), the effects of increasing water pressure at greater depths (for rebreathers in particular), and the effect of inversion. Some of this can be achieved in shallow-water exercises and with the aid of equipment such as a shallow water escape training (SWET) chair. The SWET chair partially simulates the situation in a helicopter, as the trainee is strapped into a seat with a harness, sometimes with an exit to escape through, but it falls short of simulating escape from a helicopter cabin. This will be achieved only with the inclusion of UES training. Use of EBS in the UES will provide the trainee with the skills to deploy the unit from its stowed position in the protective equipment system, see how this integrates with the seat harness, and develop skills in operating the device while escaping through an emergency exit. All of this is necessary if the individual is expected to carry out actions in the correct order and without error in a real accident. Any factor that causes uncertainty could result in a failed action and a failure to escape from the aircraft. However, the desire for high-fidelity training must be balanced against any real or perceived risks that this might entail.

There will always be a proportion of trainees who find any helicopter safety training in water stressful (Coleshaw, 2006; Harris, Coleshaw, & MacKenzie, 1996; Robinson, Sünram-Lea, Leach, & Owen-Lynch, 2008). Coleshaw (2006) reported that some trainees found EBS training (with a rebreather device) very difficult, but those who gained sufficient confidence in the use of the EBS in shallow water exercises found that it reduced any anxiety over completing the helicopter underwater escape exercises.

Tipton et al. (1997) undertook research in which trained naïve subjects used a simple rebreather and a compressed-air device in air and under water. They concluded that performance when using each device was significantly improved by in-water training, stating: "This is primarily because it gives the opportunity for individuals to get used to the combined stresses of using a new piece of equipment and performing a helicopter underwater escape" (p. 13). Bohemier et al. (1990) found that use of EBS was beneficial in helping subjects to deal with escape from more difficult seat positions in the UES requiring cross-cabin escapes, resulting in an improved probability of making a successful escape.

While some civilian user groups have initially undertaken only dry training on the introduction of an EBS product, a transition to some form of in-water training has been observed once a product has been accepted across the industry and integrated into other parts of the survival or emergency response training. There are international civilian training standards for in-water training with EBS (e.g., OPITO, 2013). While military aircrews regularly undertake UES training using compressed-air EBS there has been a reluctance to undertake passenger training using compressed air in the helicopter simulator. This to a large extent comes down to the mitigation measures that can be put in place to reduce the small barotrauma risk to a level that is considered acceptable. Mitigation may include a more rigorous medical examination and/or screening, the provision of recompression facilities, a consideration of staff–student ratios, and measures to ensure that trainees understand that they must exhale during ascent.

8.5.2 Frequency

EBS are currently used by two main occupational groups: the aircrews who fly in helicopters for a living and civilians working in the offshore oil and gas industry who use helicopters as a means of transport. The former group covers military aircrew, and civilian aircrew involved in the offshore industry. Aircrews fly on a daily basis, whereas the helicopter passengers generally spend only a few hours in the helicopter every two to three weeks. The level and frequency of training tends to reflect these differences, with military training generally refreshed every one to two years, whereas civilian offshore survival training is repeated every three to four years in many jurisdictions. Many have questioned whether civilian training is too infrequent. EBS training is generally undertaken alongside HUET, so the same frequency problem applies.

Summers (1996) investigated HUET frequency and considered that a two-year training interval was too long. She pointed out that procedural skills that are practiced infrequently decay rapidly. Mills and Muir (1999) reported that 30% of trained participants failed a HUET test undertaken six months after the training. More research is needed to determine optimum training methods and the optimum number and type of exercises required to achieve retention of knowledge and skills. This might involve additional practice during the initial learning of basic skills, more frequent further training, or methods to help retain skills during the gap between further practical training sessions such as learning conducted using electronic media.

8.5.3 Fidelity

As helicopter safety and underwater escape training have developed over the years, there have been recommendations made to increase the fidelity of the training (CAA, 2014a; Mills & Muir, 1999; Taber, 2014; Wells, 2010), to make exercises as realistic as possible (see Chapter 7 for further detail on fidelity in training). Kozey, McCabe, and Jenkins (2006) demonstrated that those who received high-fidelity HUET, with multiple inversion exercises, all undertaken with exits fitted, showed the lowest rate of failure in an inverted underwater escape test performed six months later. This research supports the intuitive view that more exercises conducted during training will produce greater levels of learning and skill acquisition when the individual is put to the test at a later date. If this is applied to EBS training, then multiple exercises should be undertaken under conditions that are as realistic as possible, as an integral part of the escape training process.

If undertaken effectively, more realistic training has obvious benefits, with the expectation that the individual will be better able to cope in a real accident. This must be balanced against any additional risks during training and additional stress that may be experienced by the trainee (see Chapters 3 and 6 for a further discussion of decision-making and coping in stressful situations). The work of Mills and Muir (1999) found that higher fidelity training caused more stress during the training, although trainees were more confident as a result of the training. There is therefore a balance to be achieved between fidelity and psychological impact, particularly when considering the fact that most EBS users will undertake emergency response training many times during their working lives, whereas only a very small number will ever experience a real helicopter water impact.

8.6 CONCLUSIONS

This chapter provides an overview of issues relating to EBS use. Since 1995, significant development of the products has taken place, while an increasing population of military and civilian users is carrying EBS in some form, particularly in colder climates. Only now is legislation being put in place to ensure that helicopter occupants are provided with EBS. This has occurred alongside improvements in training techniques and the development of the first technical standard for EBS, provided to ensure that products on the market meet minimum levels of performance. These advances are in large part due to the recognition that too many lives are lost from drowning and that EBS use is one means of potentially saving lives in helicopter accidents.

REFERENCES

AAIB (Air Accidents Investigation Branch). (1993). *Report on the accident to AS 332L Super Puma, G-TIGH, near the Cormorant 'A' platform, East Shetland Basin, on 14 March 1992.* Report no. 2/93. London: HMSO.

AAIB (Air Accidents Investigation Branch). (1997). *Report on the accident to Aerospatiale AS332L Super Puma, G-TIGK, in North Sea south west of Brae Alpha Platform on 19 January 1995.* Report no. 2/1997. London: HMSO.

AAIB (Air Accidents Investigation Branch). (2014). *Report on the accidents to Eurocopter EC225 LP Super Puma G-REDW 34 nm east of Aberdeen, Scotland on 10 May 2012 and G-CHCN 32 nm southwest of Sumburgh, Shetland Islands on 22 October.* Report no. 2/2014. London: HMSO.

AAIB/N (Aircraft Accident Investigation Board Norway). (1998). *Air accident involving Eurocopter Super Puma 332L, LN-OBP, in the North Sea on 18 January 1996, approx. 40 nm South-West of Sola, Norway.* Report 02/98. Lillestrøm: AAIB/N.

Barker, C. O., Yacavone, D. W., Borowsky, M. S., & Williamson, D. W. (May 1992). Helicopter crash survival at sea – United States Navy/Marine Corps experience 1977–1990. In *AGARD conference proceedings, Turkey.*

Barwood, M. J., Corbett, J., Coleshaw, S., Long, G., & Tipton, M. J. (2010). A comparison of the performance of emergency underwater breathing systems in cool and cold water. *Aviation, Space, and Environmental Medicine, 81*(11), 1002–1007.

Benton, P. J., Woodfine, J. D., & Westwood, P. J. (1996). Arterial gas embolism following a 1-metre ascent during helicopter escape training: a case report. *Aviation, Space, and Environmental Medicine, 67*, 63–64.

Bohemier, A., Chandler, P., & Gill, S. (1990). *Emergency breathing system as an aid to egress from a downed flooded helicopter*. Technical Report 108. Dartmouth: Canada Oil and Gas Lands Administration.

Brooks, C. J. (1989). *The human factors relating to escape and survival from helicopters ditching in water*. RTO-AG-305E. Neuilly Sur Seine: AGARD, ISBN 92-835-0522-0.

Brooks, C. J., MacDonald, C. V., Carroll, J., & Gibbs, P. N. G. (2010). Introduction of a compressed air breathing apparatus for the offshore oil and gas industry. *Aviation, Space, and Environmental Medicine, 81*, 683–687.

Brooks, C. J., MacDonald, C. V., Donati, L., & Taber, M. J. (2008). Civilian helicopter accidents into water: analysis of 46 cases, 1979–2006. *Aviation, Space, and Environmental Medicine, 79*, 935–940.

Brooks, C. J., Muir, H. C., & Gibbs, P. N. A. (March 1999). *An initial investigation of passenger evacuation from the Super Puma helicopter*. Survival Systems Report, submitted to Natural Resources Canada.

Brooks, C. J., & Tipton, M. J. (2001). *The requirements for an emergency breathing system (EBS) in over-water helicopter and fixed wing aircraft operations*. RTO-AG-341. Neuilly Sur Seine: AGARD, ISBN 92-837-1058-4.

CAA. (1995). *Report of the Review of Helicopter Offshore Safety and Survival (RHOSS)*. CAP 641. London: Civil Aviation Authority.

CAA. (2014a). *Safety review of offshore public transport helicopter operations in support of the exploitation of oil and gas*. CAP 1145. London: Civil Aviation Authority.

CAA. (2014b). *Civil Aviation Authority Safety Directive Number: SD–2014/002*. London: Civil Aviation Authority.

CEN. (2014). *EN 250:2000 Respiratory equipment – Open-circuit self-contained compressed air diving apparatus – Requirements, testing, marking*. Brussels: Comité Européen De Normalisation.

Chen, C., Muller, M., & Fogarty, K. (1993). *Rotorcraft ditchings and water related impacts that occurred from 1982 to 1989 – Phase I*. DOT/FAA/CT-92/13. Springfield, Virginia: Federal Aviation Administration.

Cheung, S., D'Eon, N., & Brooks, C. J. (2001). Breath holding ability of offshore workers inadequate to ensure escape from ditched helicopters. *Aviation, Space, and Environmental Medicine, 72*(10), 912–918.

Cheung, B., Hofer, K., Brooks, C. J., & Gibbs, P. (2000). Underwater disorientation as induced by two helicopter ditching devices. *Aviation, Space, and Environmental Medicine, 71*(9), 879–888.

Clifford, W. S. (1996). *Helicopter crashworthiness. Study I. A review of UK military and world civil helicopter water impacts over the period 1971–1992*. CAA Paper 96005. London: Civil Aviation Authority.

Coleshaw, S. R. K. (2003). *Preliminary study of the implementation and use of emergency breathing systems*. CAA Paper 2003/13. London: Civil Aviation Authority.

Coleshaw, S. R. K. (2006). *Stress levels associated with HUET: the implications of higher fidelity training using exits*. Report SC 155; prepared on behalf of OPITO, Aberdeen. Retrieved from http://www.opito.com/library/documentlibrary/huet_stress_report.pdf.

Coleshaw, S. R. K. (2012). Minimum performance requirements for helicopter emergency breathing systems. *The Journal of Ocean Technology: Extreme Survival, 7*(3), 60–73.

Coleshaw, S. R. K. (2013). *Development of a technical standard for emergency breathing systems*. CAP 1034. London: Civil Aviation Authority.

Coleshaw, S. R. K., & Howson, D. (October 1999). Escape from side-floating helicopters. In *Paper presented at the Second International Helicopter Escape Seminar/Workshop, Billingham, UK.*

EASA. (2014). *Helicopter offshore operations. Comment-Response Document (CRD) to NPA 2013-10.* European Aviation Safety Agency - Rulemaking Directorate. Retrieved from https://easa.europa.eu/system/files/dfu/CRD%202013–10.pdf.

Government of Canada. (April 22, 2015). Regulations Amending the Canadian Aviation Regulations (Parts I and VI – Offshore Operations). *Canada Gazette, Part II Official Regulations, 149*(8). Retrieved from http://www.gazette.gc.ca/rp-pr/p2/2015/2015-04-22/html/sor-dors84-eng.php.

Gray, C. G., Thalmann, E. D., & Syklawer, R. (1981). *US Coast Guard emergency underwater escape rebreather evaluation.* Panama City, Florida: Navy Experimental Diving Unit. Report No 2-81.

Harris, R. A., Coleshaw, S. R. K., & MacKenzie, I. G. (1996). *Analysing stress in offshore survival course trainees.* OTH 94 446. Sudbury: HSE Books.

Hayes, P. A. (1991). *The development and performance of underwater escape devices.* Report to SCICAD Ltd. MSC 642060/0S5.

Hayward, J. S., Hay, C., Matthews, B. R., Overweel, C. H., & Radford, D. D. (1984). Temperature effect on the human dive response in relation to cold water near drowning. *Journal of Applied Physiology, 56*(1), 202–206.

Henckes, A., Arvieux, J., Cochard, G., Jézéquel, P., & Arvieux, C. C. (2011). Hemoptysis and pneumo-mediastinum after breath-hold diving in shallow water: a case report. *Undersea & Hyperbaric Medicine, 38*(3), 213–216.

Jamieson, D. W., Armstrong, I. J., & Coleshaw, S. R. K. (2001). *Helicopter ditching research – egress from side-floating helicopters.* London: Civil Aviation Authority. CAA Paper 2001/10.

Kozey, J., McCabe, J., & Jenkins, J. (2006). The effect of different training methods on egress performance from the modular egress training simulator. In *Proceedings of the 44th Annual SAFE Symposium, Reno Nevada.*

Leach, J. (2004). Why people freeze in an emergency: temporal and cognitive constraints on survival responses. *Aviation, Space, and Environmental Medicine, 75*(6), 539–542.

Masaoka, Y., & Homma, I. (1997). Anxiety and respiratory patterns: their relationship during mental stress and physical load. *International Journal of Psychophysiology, 27*(2), 153–159.

Masaoka, Y., & Homma, I. (2001). The effect of anticipatory anxiety on breathing and metabolism in humans. *Respiratory Physiology, 128*(2), 171–177.

Mills, A. M., & Muir, H. (1999). *Development of a training standard for underwater survival.* Cranfield University Report. Prepared for Shell Aircraft.

Muir, H. (1999). Human behaviour in emergency situations. In *Proceedings of Railtrack Conference: Putting people at the centre of a safer railway, London.*

Odum, W. T. (1963). *Development of emergency breathing equipment.* Panama City, Florida: US Navy Mine Defence Laboratory. Report 197.

OLF. (2004). *Recommended OLF guidelines relating to requirement specifications for survival suits for use on the Norwegian continental shelf.* Stavanger, Norway: Norwegian Oil Industry Association.

OPITO. (2013). *Approved Standard. Basic offshore safety induction & emergency training, helicopter underwater escape training and further offshore emergency training.* Retrieved from http://www.opito.com/media/downloads/bosiet-huet-foet.pdf.

Rice, E. V., & Greear, J. F. (1973). Underwater escape from helicopters. In *Proceedings of the Survival and Flight Equipment Association Annual Symposium. Phoenix, Arizona, 1973* (pp. 59–60).

Risberg, J. (November 1997). Health risks associated with helicopter escape training. In *Paper presented at the Helicopter Escape Seminar/Workshop.* Billingham: Nutec.

Robinson, S. J., Sünram-Lea, S. I., Leach, J., & Owen-Lynch, P. J. (2008). The effects of exposure to an acute naturalistic stressor on working memory, state anxiety and salivary cortisol concentrations. *Stress, 11*(2), 115–124.

Ryack, B. L., Luria, S. M., & Smith, P. F. (1986). Surviving helicopter crashes at sea: a review of studies of underwater egress from helicopters. *Aviation, Space, and Environmental Medicine, 57*, 603–609.

Ryack, B. L., Smith, P. F., Champlin, S. M., & Noddin, E. M. (1977). *The effectiveness of escape hatch illumination as an aid to egress from a submerged helicopter: Final Report.* Report No 857. Groton, Connecticut: Naval Submarine Research Laboratory.

Shah, S., Thomas, S., & Gibb, E. (2007). Pneumomediastinum after shallow water diving. *Journal of Emergency Medicine, 36*(1), 76–77.

Sowood, P. J. (1989). *Breathing devices to aid escape from submerged helicopters; performance in cold water.* RAF IAM Report No. 584. Farnborough: Institute of Aviation Medicine.

Summers, F. (1996). *Procedural skill decay and optimal retraining periods for helicopter underwater escape training.* Willetton, Western Australia: IFAP.

Taber, M. J. (2014). Simulation fidelity and contextual interference in helicopter underwater egress training: an analysis of training and retention of egress skills. *Safety Science, 62*, 271–278.

Taber, M. J., & McCabe, J. (2009). The effect of emergency breathing systems during helicopter underwater escape training for land force troops. *Safety Science, 47*, 1129–1138.

Tipton, M. J., Balmi, P. J., Bramham, E., Maddern, T. A., & Elliot, D. H. (1995). A simple emergency underwater breathing aid for helicopter escape. *Aviation, Space, and Environmental Medicine, 66*, 206–211.

Tipton, M. J., Franks, C. M., Sage, B. A., & Redman, P. J. (1997). An examination of two emergency breathing aids for use during helicopter underwater escape. *Aviation, Space, and Environmental Medicine, 68*(10), 907–914.

Tipton, M. J., Stubbs, D. A., & Elliot, D. H. (1991). Human initial responses to immersion in cold water at three temperatures and after hyperventilation. *Journal of Applied Physiology, 70*(1), 317–322.

Tipton, M. J., & Vincent, M. J. (1989). Protection provided against the initial responses to cold immersion by a partial coverage wet suit. *Aviation, Space, and Environmental Medicine, 60*, 769–773.

TSB (Transportation Safety Board of Canada). (2010). *Main gearbox malfunction/collision with water Cougar Helicopters Inc. Sikorsky S-92a, C-GZCH St. John's, Newfoundland and Labrador, 35 Nm E 12 March 2009.* Aviation Investigation Report A09A0016. Gatineau, Quebec: Transportation Safety Board.

Wells, R. (2010). *Offshore Helicopter Safety Inquiry, Canada-Newfoundland and Labrador. Volume 1, Report and Recommendations, Phase 1.* St John's: Canada-Newfoundland and Labrador Offshore Petroleum Board. Retrieved from http://www.cnlopb.ca/pdfs/ohsi/ohsir_vol1.pdf.

CHAPTER 9

Immersion Suits for Helicopter Transportation

Dana H. Sweeney
Falck Safety Services Canada Inc., Dartmouth, NS, Canada

9.1 INTRODUCTION

Immersion suits protect wearers from cold water by reducing the intensity of cold shock (Tipton & Vincent, 1989), delaying the onset of hypothermia (Power, Simoes-Re, Barwood, Tikuisis, & Tipton, 2015; Sweeney, Ducharme, Farnworth, Prayal-Brown, & Potter, 2011), providing flotation, and increasing conspicuity, while allowing adequate mobility to perform vital actions (Canada/Canadian General Standards Board CAN/CGSB-65.16-2005, CAN/CGSB-65.17-2012; International Standards Organization (ISO) 15027-1). Immersion suits can be broadly categorized as emergency abandonment immersion suits (AISs) or constant-wear suits. AISs are intended to be readily available and rapidly donned in an emergency, whereas constant-wear suits are worn proactively in situations in which an emergency would not afford the time and/or space to don the immersion suit. Helicopter transportation suit (HTS) systems are constant-wear immersion suits that: (1) need to offer appropriate mobility and comfort during normal flight conditions, (2) are designed to facilitate underwater egress (i.e., be minimally buoyant during egress), (3) must provide adequate buoyancy after reaching the water surface, (4) must demonstrate effective integration with survival equipment and accessories, and (5) offer appropriate thermal protection for cold-water immersion.

In areas where helicopters fly over warm water, such as the Gulf of Mexico, personal protective equipment for flights typically consists of regular clothing and an inflatable life jacket, as required by the Federal Aviation Regulations (FARs). Where helicopter operations occur over cold water (e.g., Canada, North Sea, Norway, Southern Australia, and the Arctic regions), constant-wear HTS systems are regulated to be worn by passengers and pilots/aircrew.

When HTSs are required for the helicopter flight, pilots and passengers typically wear different suits. During flight, pilots are busy flying, require

Handbook of Offshore Helicopter Transport Safety
ISBN 978-1-78242-187-0
http://dx.doi.org/10.1016/B978-1-78242-187-0.00009-2

more mobility, and generate slightly more metabolic heat (approximately 1.5 metabolic equivalents (Ducharme, 2006; Faerevik & Reinertsen, 1998), which is just slightly above resting levels of exertion). Additionally, pilots experience intense solar energy through the large cockpit windows (Gaul & Mekjavic, 1987; see also Taber, 2010, for a discussion of the greenhouse effect). Offshore (oil and gas industry) passengers remain sedentary, and they are not exposed to the same thermal stressors as the pilots. Although many of the concepts covered in this chapter apply equally to pilots and passengers, the majority of the content is focused on passenger HTSs.

Many aspects of immersion suit design and performance apply to all immersion suit types. Readers are directed to Sweeney and Taber (2014) for a discussion on immersion suit design considerations that apply to all dry suits for cold-water immersion protection. As an extension of that work, this chapter explores HTS system design and standards that establish the criteria for normal and emergency use of immersion suits in helicopters. It must be noted that the important issue of the required maintenance associated with the frequent use of HTS systems is not discussed in this chapter, but proper maintenance is crucial for ensuring thermal protection during cold-water immersion.

9.2 HELICOPTER IMMERSION SUITS

National and international aviation regulations, such as the Canadian Aviation Regulations (CARs) and Joint Aviation Regulations (JARs), mandate when passengers must wear an HTS. CARs require passengers to wear an HTS if: (1) the flight occurs over water less than 10 °C and (2) the flight plan involves travel over water that would require the helicopter to carry life rafts (CARs 602.63 (Parts 4 and 5)). Similarly, JAR-OPS 3.837 (Part 1) requires that passengers must wear an HTS for helicopter travel related to offshore operations in hostile sea areas more than 10 min from land if: (1) the water temperature is less than 10 °C, (2) the estimated survival time for immersed persons (presumably unprotected) is less than the estimated rescue time, or (3) the flight occurs at night.

Each year, there are hundreds of thousands of helicopter flights globally (available data for the Gulf of Mexico and North Sea offshore regions indicate approximately 145,000 flights per year for those two regions alone) (Helicopter Safety Advisory Conference, 2014; Oil & Gas, 2011), many of which (those in the North Sea) would meet the JAR-OPS requirements for the use of an HTS. Using data gathered from offshore oil operations, Taber (2014) reports that between 2000 and 2012 there were 98 helicopter ditching/water impact events recorded. Based on the locations of these ditching incidents, approximately 25% of passengers involved would have been wearing an HTS. Given the number of flights that would require the use of HTSs, and the low incidence rate of ditching, the majority of HTS use occurs in normal flight. The wearing of an HTS in flight to protect against potential cold-water immersion introduces two critical design dichotomies: (1) providing adequate thermal protection while providing minimal suit buoyancy during egress and (2) achieving the proper balance between thermal comfort during routine helicopter transportation and sufficient thermal protection in the event of cold-water immersion.

The following sections discuss these specific design dichotomies and challenges of these competing priorities in HTS design. The discussion begins with specific aspects of suit design, use, and testing, followed by issues of operational and practical significance.

9.2.1 HTS Configurations and Donning

AISs need to have simple sizing systems and donning procedures to facilitate rapid emergency distribution and donning (Sweeney & Taber, 2014). Theoretically, HTSs do not have these same donning constraints because they are donned proactively before flight. As a result, CAN/CGSB-65.17-2012 does not have a donning time requirement for HTSs, whereas under ISO 15027-1, HTSs are classified as constant-wear immersion suits that could also be donned in an emergency. Therefore ISO 15027-1-approved HTSs have a 2-min donning requirement (including a life jacket). If HTSs are required for flight, donning occurs in a controlled environment in which assistance is available, never in an emergency situation. Figure 9.1 shows an example of an HTS with a front-entry zipper configuration and a face seal. The controlled, predeparture environment alleviates some of the need for donning simplicity that is critical in an AIS (Leese & Norman, 1979; Sweeney & Taber, 2014).

9.2.2 Neck Seals

Sweeney and Potter (2010) demonstrated that immersion suits with properly fitting neck seals permitted nearly 15 times less water ingress than appropriately fitted immersion suits with face seals during immersions with challenging wind and wave conditions. Although neck seals offer superior

Figure 9.1 Insulated HTS with front-entry zipper and face seal. This suit has integrated boots and inflatable buoyancy elements. Hand protection is stored in the arm pockets.

protection against water ingress they are also associated with more complicated donning procedures. The various donning configurations used with HTSs are adapted according to regulations and client needs that improve suit fit, reduce bulk, and ensure effective equipment integration into the suit system.

The use of neck seals allows zipper configurations such as diagonal (Figure 9.2(a) and (b)) or horseshoe (Figure 9.3(a) and (b)) entry, and other configurations are also possible. With these zipper configurations, donning the suit involves pulling the neck seal over the head and sealing the zipper to achieve watertight integrity (assuming any other seals, such as wrist seals, are properly fitted). Although many passenger HTSs utilize a neck seal with a front diagonal zipper, pilot and aircrew HTSs often use the other zipper entry styles to facilitate the integration of specific equipment and operational requirements.

9.2.3 Insulation Configurations

To reduce heat loss during cold-water immersion, HTSs normally have insulation provided by an integrated insulated liner or layers of clothing. The latter configuration typically provides less buoyancy than an HTS with inherently buoyant insulation material, and the majority of buoyancy comes from an inflatable buoyancy element.

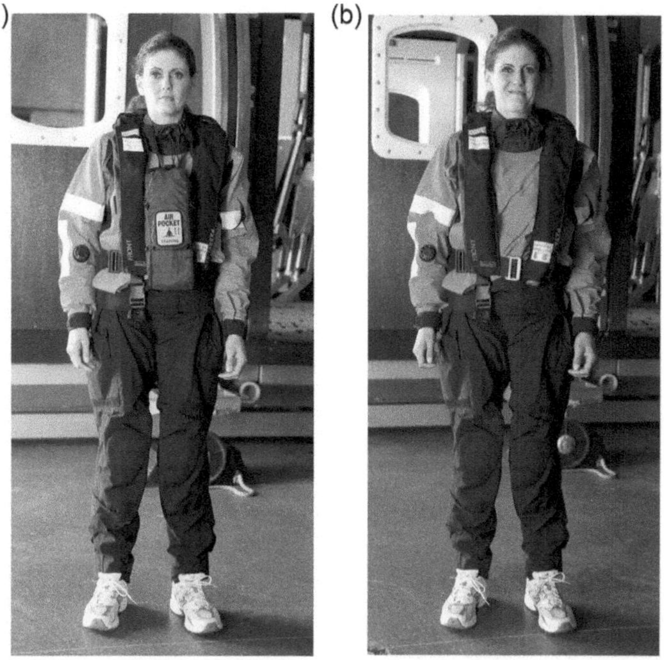

Figure 9.2 (a) Representative uninsulated shell HTS with neck seal and wrist seals. With this configuration, the hood would probably be stored in the thigh pocket and there would be gloves attached to the arms. External footwear is required over the waterproof socks. The suit is fitted with a separate Technical Standard Order-C13f-approved aviation life jacket and rebreather emergency breathing system (EBS). (b) EBS removed to show the diagonal waterproof entry zipper crossing from the right hip to the left shoulder.

An advantage of a design that depends on layers of clothing is that the system is more adaptable to environmental conditions; adjusting the amount of insulation is achieved by altering undergarment requirements. For example, the UK sector of the North Sea specifies that two layers of clothing are required in summer months when the water is above 10 °C, and in colder months, passengers are required to wear three layers of clothing underneath the suit shell (Figure 9.4).

For HTSs with integrated insulation, different insulated liners could be introduced for summer and winter seasons, if the regional regulatory system permits these modifications. Some standards, such as CAN/CGSB-65.17-2012, have no allowances for optional suit-liner configurations. With no options for the amount of insulation, all HTSs approved in Canada must provide a minimum of 0.75 immersed clo (Section 5.22.2.1), which

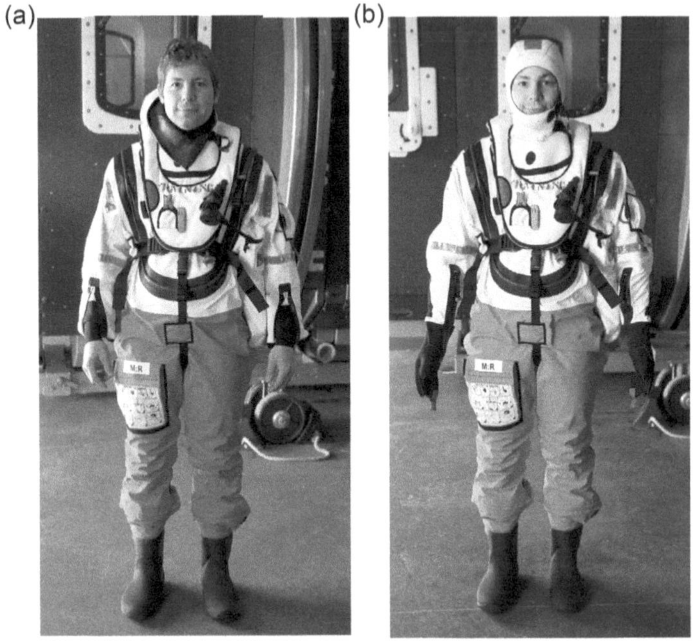

Figure 9.3 (a) Insulated HTS with horseshoe-style entry and a neck seal. An attached hood also creates a secondary face seal. Zippered overmitts (see Sweeney & Taber, 2014 for more discussion) provide hand protection. Life jacket is integrated but secured with external straps. (b) Hood and zippered overmitts in donned configuration.

increases the minimum buoyancy of the HTS. In comparison, ISO 15027-1 has four classifications of suit insulation (see Table 9.1), providing the thermal protection benchmarks for seasonal insulation changes.

9.3 SEALING SYSTEMS AND ACCESSORIES

Once the suit is donned and the majority of the body is protected, head and hand protection is an important consideration. Hoods that are used with neck-seal suits can be either attached or detached. When hoods are detached, they are often stored in a pocket and are donned prior to, or following, egress. When hoods are attached to the HTS, they may be designed to be removed from the head during flight, but remain in close proximity for rapid redonning in the event of an emergency.

If passengers are permitted to unseal the suit during flight, HTS systems must be able to be rapidly resecured to a watertight condition. Some suit styles are able to be unzipped and have the hood removed during flight.

Figure 9.4 Information provided to helicopter passengers in the UK offshore oil and gas sector about acceptable clothing for wear under HTSs and required layers for summer and winter seasons. *(Image from Step Change in Safety (2014b)—https://www.stepchangeinsafety.net/safety-resources/helicopter-safety/helicopter-safety-resources/standard-clothing-policy.)*

Table 9.1 ISO 15027-1 Specifies Four Thermal Performance Levels for Immersion Suits. The Expected Time (in hours) for Deep Body-Core Temperature to Decrease below 34 °C (at Various Water Temperatures) for Each Classification (A, B, C, and D) is Shown

	Insulation Classification			
	A	B	C	D
Water Temperature °C	**0.75 clo**	**0.50 clo**	**0.33 clo**	**0.20 clo**
Less than 5	6 h	2.5 h	1.5 h	1 h
Between 5 and 10	9 h	4.5 h	2.5 h	1.5 h
Between 10 and 15	15 h	7 h	4 h	2 h
Above 15	24 h	15.5 h	6 h	3 h

To maintain the thermal protection properties for water immersion, it is critical that watertight integrity can be achieved quickly if a ditching is imminent. When a manufacturer specifies that the HTS is able to be unsealed during flight, the HTS approval process in accordance with CAN/CGSB-65.17-2012 and ISO 15027 requires demonstration that the suit can be completely resealed in less than 10 s. This demonstration must be completed while seated and secured with an appropriate seat belt. Based on manufacturer specifications, some HTS designs may not permit unsealing during flight, even if the sealing system is accessible. For review of issues concerning hand protection, see Brooks (2003) or Sweeney and Taber (2014).

9.3.1 Suit Fit

Whereas AIS systems generally follow the small–universal–jumbo size model to facilitate emergency distribution (Sweeney & Taber, 2014), HTSs are provided with discrete size ranges (e.g., XS, S, M, L, XL). It is normal practice for HTS systems to be available in 10–15 or more sizes to ensure the best fit for wearers. Having a proper-fitting HTS is important for performance for the following reasons: (1) the amount of air that can be trapped within the suit is minimized (Leese & Norman, 1979; and discussed in more detail below under escape buoyancy); (2) overall bulkiness of the suit is reduced, which improves mobility and survival equipment integration; and (3) water ingress is minimized with properly fitted seals.

Other approaches to achieving proper suit fit may include designs such as internal suspension/tensioning systems that allow for more customized fit (Figure 9.5), elasticized sections that allow the suit to stretch and

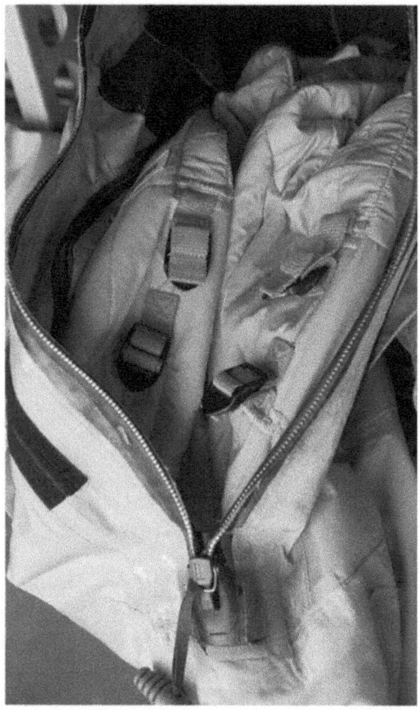

Figure 9.5 Example of internal suspension straps fitted inside an insulated HTS. This system allows suit fit to be adjusted and also assists with keeping the boots in position.

accommodate a larger size range and also improve mobility, and external compression systems that greatly reduce the amount of air that can be trapped within the suit (Figure 9.6).

9.3.2 Escape Buoyancy

The potential impact of buoyancy on underwater egress is recognized by all major HTS standards, and therefore escape buoyancy (EB) is a design constraint on HTS systems. EB is defined by CAN/CGSB-65.17-2012 as "the buoyancy of the suit system that must be overcome when escaping from an immersed, inverted helicopter. Note: it includes the inherent buoyancy of the components of the suit, and entrapped air, but excludes the inflated buoyancy of an inflatable buoyancy element when fitted" (p. 3). The CAN/CGSB-65.17-2012 definition of EB excludes upright submersions; however, if a helicopter submerges without inverting, HTS buoyancy will act to carry the occupant to the ceiling of the helicopter, whereas if the helicopter inverts, buoyancy will tend to hold the occupant in his or her seat. In the event of a helicopter ditching and inversion/ submersion, excessive buoyancy may increase the difficulty of egress

Figure 9.6 Example of elasticized material on the exterior of suit. This technique allows the shell to be oversized to accommodate a larger body size range. The elastic material compresses the extra material to maintain minimum internal suit volume and trapped air.

(Brooks, 1988, 1989; Brooks & Provencher, 1984; Kozey, Potter, & Sweeney, 2011; Leese & Norman, 1979). As evidence of this possibility, Brooks, Muir, and Gibbs (2001) reported that suit buoyancy was the most common issue that complicated underwater egress from a Modular Egress Training Simulator (METS™).

Learning to control the buoyancy through physical references and movement techniques is an important aspect of helicopter underwater egress training (HUET) (Bohemier, Brooks, Morton, & Swain, 1998; Brooks, 1989). Helicopter underwater egress may be influenced by other physical elements such as water motion, seat position (crash-attenuated or not) (Taber, 2013), egress movements (Coleshaw, 2006), distance to an exit (Brooks & Bohemier, 1997; Taber, 2014), seat-belt release mechanism (Brooks, 1989), required window jettison actions (Brooks & Bohemier, 1997), position of the helicopter (Taber & McCabe, 2007), water temperature, and time of day (Taber & McCabe, 2006) (see Chapter 7). When combined with the challenges of the physical environment, excessive EB could further complicate the action required for successful egress. Although the effect of total buoyancy on underwater egress is an obvious concern, few studies (Brooks, 1988; Brooks & Provencher, 1984; Kozey et al., 2011) have specifically investigated this effect. Another relevant study by Gagnon, McDonald, Pretorius, and Giesbrecht (2012) investigated the effect of buoyancy on underwater vehicle egress. It was found that egress difficulty increased with buoyancy; however, the effect did not become significant until underwater egresses were completed with a 156-N inflated life jacket (see also McDonald & Giesbrecht, 2013). Table 9.2 lists conversions for buoyancy amounts common in survival equipment.

The first study to investigate the effect of buoyancy on underwater egress was conducted by Brooks and Provencher (1984). In this study, participants completed simulated underwater egress activities while wearing various levels of buoyancy, to a maximum of 267 N. The egress activities required trained divers to move 2.4 m through a flooded chamber from a chair to a window opening. Divers and nondivers completed breath-hold dives to 3 m, retrieved a 0.5-kg mass, moved through a window opening, retrieved a second 0.5-kg mass, and returned to the surface. Brooks and Provencher (1984) purported that divers began reporting difficulties completing these tasks when they were wearing between 120 and 191 N of buoyancy, and some were unable to complete the tasks when wearing between 160 and 254 N of buoyancy. In the 3-m diving activity, the divers had difficulty completing the task when wearing between 133 and 218 N of buoyancy, with failures occurring at buoyancy levels between 173 and 267 N. Nondivers reported difficulty in completing the pool diving activity when wearing between 76 and 138 N, with failures occurring at buoyancy

Table 9.2 Conversion from Newtons to Pounds of Force of Common Buoyancy Values Used for Marine and Aviation Survival Equipment

Newtons (N)	Pounds of Force (lbf)
4.4482	1.00
70	15.74
100	22.48
147	33.05
150	33.72
156	35.07
175	39.34
275	61.82

Table 9.3 Summary of Results for Pool Diving Activity

Population–Activity	Difficulties	Failures
Divers–flooded chamber	120–191 N	160–254 N
Divers–3 m diving	131–218 N	173–267 N
Nondivers–3 m diving	76–138 N	98–178 N

Data from Brooks and Provencher (1984).

levels between 98 and 178 N. The results of Brooks and Provencher (1984) are summarized in Table 9.3.

The results of the Brooks and Provencher (1984) study provided the first information on the effect that buoyancy has on the ability to perform underwater activities. Brooks (1988) improved the specificity of the research by directly investigating the effect of buoyancy during inverted helicopter underwater escape. The results of this second research study led to the establishment of a maximal buoyancy, which HTS systems could not exceed, for the Canadian CAN/CGSB-65.17-M88 standard. Brooks (1988) had participants conduct underwater egress trials in an underwater egress simulator (UES) while wearing various amounts of buoyancy. Based on the results it was determined that participants began to have significant difficulty in completing egresses when wearing approximately 147 N (33 lb) of buoyancy. In 1999, a revised Canadian HTS standard (CAN/CGSB-65.17-99) raised the EB limit to 175 N (Brooks, 2003).

Before 1999, the minimum buoyancy requirement in Canada for an AIS was 156 N (CAN/CGSB-65.16-M89) and the EB limit for an HTS was 147 N, making it impossible for a single suit to achieve both approvals (CORD, 1995). After verifying the safety of changing the minimum AIS buoyancy to 70 N (Potter, O'Neill, & Brooks, 2003), the change was implemented in CAN/CGSB-65.16-2005, further increasing the opportunity for a single suit to meet both the Canadian AIS and HTS standards. By 2005, there was a 105 N gap between the AIS and the HTS minimum and maximum buoyancy limits in the Canadian standards.

No additional research to evaluate the impact of the increased EB was completed until Kozey et al. (2011) conducted a three-phase study to further understand the relationship between buoyancy and underwater egress performance. These three phases of research represent the most thorough and specific investigation of EB effects on helicopter underwater egress completed to date.

The first phase of the EB study by Kozey et al. (2011) involved total buoyancy measurements on 41 participants (32 males and 9 females) immersed in an upright, seated position while wearing HTS systems (CAN/CGSB-65.17-99 method). Each participant was tested in two different HTSs, one that was properly fitted and one that was oversized by one size. To explore realistic scenarios, each test condition was measured with all air exhausted and also in a "just sealed" condition, where no air had been purposefully exhausted from the suit after donning. The oversized trials represented the standard fitting procedures for HTSs in Canada, which were based on visual estimation of height and weight (TSB, 2010). This procedure permitted passengers to "upsize" their suits for mobility (TSB, 2010). It was found that across a wide range of body sizes, wearing an appropriately sized HTS reduced the total buoyancy by 7 ± 23 N, compared to an oversized suit. The practice of upsizing is no longer possible in Canada, as all passengers are fitted for an HTS using manufacturer-specified criteria and tests (such as the Hansen Protection HTS-1 fitting guide). Once a passenger is fitted for an HTS, only that size is issued to him or her for flight.

In the second phase, 53 participants (40 males and 13 females) were evenly grouped by body size, and each group was assigned a different amount of buoyancy (111, 134, 156, 178, 201, and 223 N). Participants completed three underwater egress-related tasks: (1) submerge in shallow water, turn a handle, and move through a helicopter window replica; (2) dive to a depth of 1.8 m to retrieve an underwater target; and (3) complete an egress task while inverted and move underwater along a guide rail for 3.5 m to an exit window to egress the METS™.

Results indicated that there were three handle-turning failures and four failures to move through the underwater opening in task 1. In task 2, 70% of participants were unable to retrieve the buoyant target from a depth of 1.8 m. In task 3, the 111 N group took significantly less time than the 201 N group (10 and 15 s, respectively) to complete the egress task. Although there was a significant increase in time required for egress, there was no difference in the pass/fail frequency with increasing buoyancy. It was noted that any increase in time required to complete an underwater task could be a fatal, especially in cold water, in which breath-hold times are greatly reduced (Barwood, Dalzell, Datta, Thelwell, & Tipton, 2006; Jay & White, 2006; Taber, MacKinnon, Power, & Walker, 2015). From these results, it was determined that buoyancy greater than 178 N was problematic for egress performance owing to the increase in time required to egress. Also, passengers are unable to free dive to reach exits even when wearing low amounts of additional buoyancy.

For the third phase, 48 participants (39 males and 9 females) completed a series of egress maneuvers while inverted in a METS™. In this final phase, participants' body sizes were balanced across the three groups. Group participants were assigned a level of buoyancy, either 134, 178, or 223 N. Interestingly, no significant differences in egress time were recorded. Two small females from the 223 N buoyancy group withdrew owing to excessive buoyancy that prevented them from performing the necessary actions in the UES. Their voluntary withdrawal from testing suggests that at some point buoyancy creates egress problems that are too difficult, or panic inducing, to allow completion of a successful egress. Fortunately, it is unlikely that small passengers would ever have an HTS that had 223 N of buoyancy (excluding the possibility of an accidental life jacket inflation).

There are a number of other issues that must be considered when evaluating the phase 3 results of Kozey et al. (2011). This testing was completed in a calm pool, and shortly before completing the egress tasks, the participants were carefully instructed on the techniques for underwater, inverted movement and exit operation. As mentioned above, many physical conditions could exist that impede egress after a ditching. Such conditions could greatly increase the difficulty of controlling higher amounts of buoyancy and precipitate the loss of physical references, thus decreasing the ability to find or reach available exits.

9.4 EGRESS SKILL RETENTION AND EB

Mills and Muir (1999) reported degradation of the accuracy and efficiency of underwater movements after an interval as short as 6 months. Brooks and Provencher (1984) found that trained divers were better at completing egress tasks than nondivers, suggesting that the diver training and experience working in the water while wearing buoyant protective equipment was beneficial for completing the required underwater tasks. With infrequent exposures to the underwater environment while wearing buoyancy, it is reasonable to expect that any proficiency that develops for managing HTS buoyancy will degrade similar to the skill degradations reported by Mills and Muir (1999). Interestingly, the CAN/CGSB-65.17-2012 standard requires that participants have no previous experience with HTS and have not had HUET for at least 2 years, whereas recurrent HUET requirements for offshore workers is 3 years in Canada (CAPP, 2013) and 4 years under Offshore Petroleum Industry Training Organization standards (OPITO, 2013), for example (see Chapter 2).

9.5 EB TEST METHODS

Major HTS standards such as the CAN/CGSB-65.17, European Aviation Safety Agency/European Technical Standard Order (EASA/ETSO) 2C503, ISO 15027-1, and the Norwegian 094 all address EB. However, given that the Norwegian 094 and EASA/ETSO 2C503 both specify the EB test methods outlined in the ISO 15027-3 standard, there are actually only two EB test methods in practice.

9.5.1 Canadian Test Standards

Since the first edition of CAN/CGSB-65.17 in 1988, EB has always been rigorously measured in Canada with test methods that have increased the level of helicopter ditching realism as the versions of the standard have progressed. The test method for EB used in the early versions of CAN/CGSB-65.17 (M88 and 99) involved measurement of buoyancy during upright, seated 15-s submersions. Prior to each submersion, the suit was opened and resealed, and the participant did not enter the water. This was to simulate that they had just donned the suit and boarded the helicopter or opened the seals during flight. During the 15-s submersion, participants spent 5 s in the brace position and then 10 s seated upright, permitting some trapped air to evacuate the suit (a procedure that was written when two-point lap belts were common in helicopters). The results of the upright 15-s submersion procedure were found to correlate reasonably well with buoyancy measurements recorded in an inverted, submerged position (Brooks, 1988).

9.5.2 ISO Test Standards

The ISO 15027-3 (Paragraph 3.11.7.2) test method involves submerging test participants vertically for 15 s while they are wearing standard test clothing and the HTS system, but not the inflatable life jacket. This is problematic, as the life jacket can trap air under its valise, and the harness can restrict air movement within the suit, possibly reducing the ability for air to escape. If the suit systems do not have automatic relief valves, they must be manually vented (burped) prior to measurement. Burping the suit is typically accomplished by opening the main neck or face seal while squatting and squeezing out as much air as possible and closing the seal. The ISO 15027 EB test methods assume that no air would be accidentally or purposefully introduced into the suit after donning. However, it is not uncommon for passengers to pull the neck seal away from their neck for temporary comfort while wearing a suit. Last, by testing the person in a vertical position, there is no restriction to air movement within the suit, as it can occur while seated and wearing a seat belt. The maximum buoyancy of the ISO-approved system cannot exceed 150 N.

9.5.3 Real-World Implications of EB

After a fatal helicopter ditching in Atlantic Canada in 2009 (TSB, 2010), Canada chose to rewrite the EB test method to involve inversion in a UES. This requirement was based on the fact that 45–70% of helicopter ditchings end in immediate rollover (Brooks, MacDonald, Donati, & Taber, 2008; Taber, 2014; Taber & McCabe, 2006). When EB is tested in the upright seated or vertical position, trapped air moves to the neck area of the suit, where it can escape for 15 s through face and neck seals or valves located in the upper portions of the suit. Measuring EB while secured in a seated brace position with a four-point seat-belt harness after a full inversion creates the most realistic condition possible for measuring EB (Figure 9.7).

During the initial seconds of a dynamic inversion, air can be forced out of the upper portions of the suit, but once the wearer has passed through

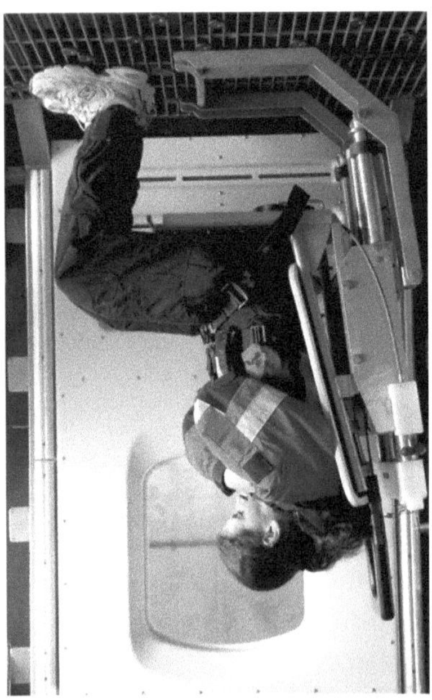

Figure 9.7 Inverted four-point brace position used during EB testing in accordance with CAN/CGSB-65.17-2012. During EB testing, the brace position is maintained for 5 s and then relaxed.

90° of roll, the air begins to move toward the feet, where it remains trapped unless the air can be evacuated by another method. It is possible that the inclusion of the inverted EB test in a national standard will lead to innovative changes in the systems used in HTS designs to eliminate or reduce the presence of trapped air while inverted.

9.6 ADDITIONAL EB IMPLICATIONS

9.6.1 Thermal Protection

In addition to buoyancy, EB also challenges other aspects of HTS design, such as thermal protection. The efficacy of insulating materials is dependent on their ability to trap air, which competes directly with the objective of minimizing escape buoyancy. Depending on the thermal protection requirements of the referenced HTS standard, providing sufficient insulation and thermal protection while achieving the EB requirements can provide a considerable challenge to suit designers. The nearly perfect buoyancy situation for helicopter passengers would be to have minimal positive buoyancy during egress (Leese & Norman, 1979), followed by the ability to have high levels of insulation and buoyancy following egress.

The required thermal protection for a suit is determined by the air and water temperatures of the operational environment and the expected time to rescue. Achieving the EB standard in a suit intended to prevent core temperature from decreasing more than 2 °C after 6 h of immersion in 0–2 °C water (ISO 15027 Class A requirement) will be more difficult than in a suit that requires less thermal protection. For example, the EASA/ETSO 2C503 standard requires that helicopter suits provide at least ISO 15027 Class B protection, whereas CAN/CGSB-65.17-2012 requires the equivalent of ISO 15027 Class A. Even though it is more challenging to keep EB minimized in the more highly insulated suit, the EB threshold is higher in the Canadian standard, which alleviates some of that challenge.

9.6.2 Water Ingress

Another major issue for all immersion protection apparel is water ingress (Sweeney & Taber, 2014). With the reduction in insulation that occurs in an immersion suit system as water enters and displaces the air from the insulating materials (Allen, 1984; CORD, 1987; Tipton & Balmi, 1996), it becomes more difficult to meet EB restrictions because more insulation material is required to meet minimum thermal requirements. When

Figure 9.8 Brace position recommended by Brooks (1989) for use in helicopters with two-point harnesses. This brace position was used during the EB testing by Brooks (1988).

clothing gets saturated with water, the air trapped within the fibers is displaced and insulation is greatly reduced (CORD, 1987; Hall & Polte, 1956; Light, Avery, & Grieve, 1987).

Although neck seals are effective at keeping water out of the HTS (Sweeney & Potter, 2010; Sweeney, Potter, & Durnford, 2008), they do not let air escape from the suit as effectively as a face seal, especially in the upright four-point brace position. Also, if a hood that covers the neck seal is being worn, air is prevented from passing out through the neck seal. Brooks (1988) observed that in the two-point brace position (Figure 9.8), the neck seal was stretched at the back of the neck, and when the internal air pressure increased during submersion, the air was forced out through the uncovered neck seal.

9.6.3 Equipment Integration and Snagging

HTS systems are typically fitted with accessories to provide the wearer with survival resources should they become immersed and/or submerged in the water. A relatively standard set of HTS survival accessories includes life jacket, gloves, hood (if required), emergency breathing system, personal

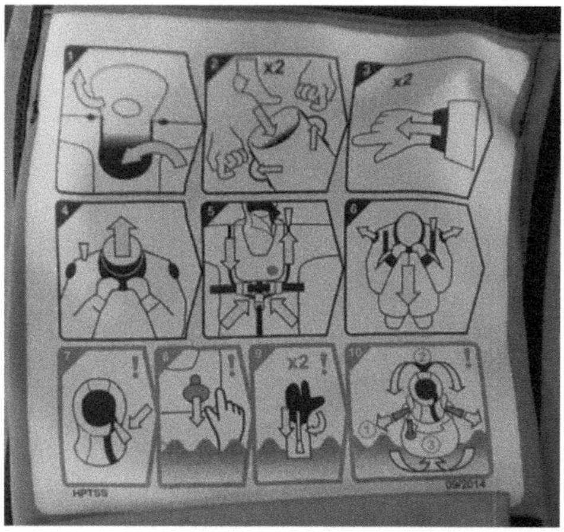

Figure 9.9 Example of pictorial HTS instructions.

locator beacon, strobe light, buddy line, whistle, and spray shield. Other accessories may include goggles and/or nose plugs. A key consideration for the design of HTS systems is that accessories are visible to the wearer both in air and in the water (with labeling that is easily read under dimly lit conditions) (Figure 9.9), accessible (preferably by either hand), and designed so that they can be used with reduced dexterity that results from gloves or cold hands (MacKinnon & Mallam, 2010). It is important that an HTS system is designed so that the suit and accessories create minimal snagging hazards during egress. All accessories must be adequately secured and be positioned to minimize bulk and preserve mobility.

9.6.4 Buddy Line

Buddy lines on immersion suits and HTSs have traditionally used snap hooks at the end of a cord. However, if the wearer were able to clip to an object with the snap hook, it may be nearly impossible to release the buddy line, especially in rough seas with cold hands or insulated gloves. In Canada, the requirements for the buddy line were revised in the CAN/CGSB-65.17-2012 standard, and it was made mandatory for the buddy line to have an easily operated quick-release mechanism within reach of the wearer so that it could be quickly released, even if the snap hook could not be reached or released (Figure 9.10). In comparison, CAN/CGSB-65.16-2005 specifies a 1- to 2-m line with a snap hook and a breaking strength between 400 and 1340 N, and ISO 15027-1 specifies a breaking strength

Figure 9.10 Buddy line quick-disconnect system to allow the buddy line to be disconnected from the attachment point while under tension or if the snap hook cannot be operated.

requirement of only 750 N. With such high breaking strengths required by suits approved to these other standards, it is impossible to release the buddy line unless the snap hook can be successfully unfastened.

9.6.5 Emergency Breathing Systems

Emergency breathing systems (compressed air and rebreather systems) should be affixed to the suit such that their mouthpieces are readily accessible in an emergency (see Chapter 8). Pressure gauges and on/off knobs should be easily identified, and their overall profile should be minimized so that the system will be unlikely to snag on protruding items in the helicopter cabin or during egress through an exit.

9.7 TEST METHODS FOR SNAGGING

The ISO 15027-3, CAN/CGSB-65.17-2012, and EASA/ETSO 2C503 standards include tests for HTSs specifically to identify potential snagging concerns that could impede egress. The CAN/CGSB-65.17-2012 snagging test requires that each test subject, while being observed by a qualified HUET instructor, complete three underwater egresses through a 460 by 550-mm window (the size of a Sikorsky S92 push-out passenger window (p. 50, TSB (2010)) from an aisle seat. The three HUET inversions are completed at approximately 45, 180, and 225° of rotation in order to evaluate if any of the suit components become dislodged or create a snagging hazard at various angles of inversion. By inverting to various angles, which could occur with a damaged flotation system, or if a side floating system was employed (Jamieson, Armstrong, & Coleshaw, 2001), the tendency for accessories to dislodge is assessed as the person moves through the exit in different orientations. The rolls are video recorded and observed by competent HUET instructors, and each participant also responds to a questionnaire after each roll so that any snag or impediment to egress can be identified.

Comparatively, the snagging test in ISO 15027-1 requires one large participant, preferably with a 95th percentile bi-deltoid width (the population for this anthropometric dimension is not defined; however, it would be preferable for the measurement to be based on a population representative of the region and workforce), to complete a single egress with all suit accessories attached and stowed. The egress is to be completed through an escape exit of 420 by 660 mm, in air as well as 300 mm under water. During these tests, the suit is not to offer any significant snagging hazards. EASA approvals require that all test participants must pass the ISO 15027-3 test method, except that they must egress through a window that is only 430 by 355 mm.

The exit sizes defined for snagging tests both by CAN/CGSB-65.17-2012 and ISO 15027-3 and by EASA for HTS approval are smaller than Class IV exits (as defined by airworthiness regulations—see Chapter 7 for further information regarding physical fidelity), although there is no explanation as to why these exit sizes were selected. EASA also specifies that at least one test participant must have a shoulder width of 500 mm, a requirement that is currently outdated, since the Civil Aviation Authority no longer permits passengers with a shoulder breadth greater than 559 mm (considered the extra broad passenger category) to sit next to an emergency exit smaller than Class IV (CAA, 2014; Step Change in Safety, 2014).

9.8 IMPORTANCE OF THE STANDARD

As described in Sweeney and Taber (2014), suit design is guided by many variables and performance requirements, including the suit standard referenced by regulations. Suit standards define the minimum performance requirements and the methods of determining those levels of performance. These criteria are based on the collective input of the stakeholders concerned with the safety of helicopter passengers (regulators, industry, manufacturers, and end users). Through effective standard writing techniques (ISO, CGSB, ETSO, etc.), testing methods become a balanced set of performance criteria that reflect the operational requirements and the safety philosophies of the individual stakeholders. This process has a critical impact on the suits that are developed to a particular standard, as the test methods for each aspect of performance guide the design and establish the resulting performance.

For example, if an HTS system were developed to pass the EB test of CAN/CGSB-65.17-99, and was then subjected to the test methods of CAN/CGSB-65.17-2012, very different levels of suit performance would probably be observed. Designing an HTS to have venting characteristics for a 15-s upright, seated immersion will not lead to a venting system that functions well during an inverted, seated immersion. The reverse is also true. This example demonstrates that when evaluating the performance of any immersion suit system, it is critically important to understand the standard test methods used when the system was approved.

9.9 THERMAL COMFORT AND SITUATIONAL AWARENESS

The high amount of insulation necessary to provide adequate thermal protection during cold-water immersion increases heat stress during flight, and Faerevik and Reinertsen (1998) reported that when wearing an HTS, skin temperature began to increase when cabin temperature was only 18.9 °C. As skin temperature increases and sweating begins, the humidity inside the suit increases. Inside the suit microenvironment, there is very little dry heat transfer (conduction, convection, and radiation) occurring, as these heat transfer pathways depend on the existence of a temperature gradient from the warmer skin to the cooler surrounding environment. Therefore, inside the suit, the only avenue for heat loss is evaporation. Evaporation requires the surrounding environment to have a lower vapor pressure (less humid) than the skin. As moisture levels inside the suit increase, all heat loss pathways are essentially ineffective, and a situation known as uncompensable heat stress is established (Cheung, McLellan, & Tenaglia, 2000). In an effort to alleviate the uncompensable heat stress environment (Ducharme, 2006; Faerevik & Reinertsen, 1998; Taber, Dies, & Cheung, 2011), many HTS systems use waterproof breathable fabrics to allow some evaporative cooling to occur and keep the head exposed with removable hoods that can be donned in an emergency. Also, some HTS models are able to be unzipped during flight for ventilation.

Waterproof breathable fabrics are reported to have vapor transmission rates of less than 200 g/h/m^2 (Holmes, 2000), whereas resting sweat rates can be as high as 246 g/h/m^2 (excluding head and hands, which are typically not enclosed within the body of the HTS—sweat rates were calculated from the results of Taylor & Machado-Moreira, 2013). Additionally, Sullivan, Mekjavik, and Kakitsuba (1987) reported air exchange out of the HTS to be as low as 30 mL/min (1.8 L/h) with waterproof breathable fabrics, which does not provide sufficient ventilation for substantial heat and moisture removal. This disparity supports the likelihood of a net gain of moisture inside the HTS. Compounding the problem, when impermeable materials (e.g., foam) are included in the undergarment system, water vapor movement to the breathable membrane will be further slowed. Regardless, in the survey study by Gaul and Mekjavic (1987) respondents reported that breathable fabrics in HTS systems significantly improved thermal comfort ratings, decreased sweating, and decreased wetting of insulation materials.

Helicopter passengers do not have the same physical requirements as aircrew. Passengers are passive occupants during flight, unless an emergency situation develops and they must begin preparations for emergency egress. Research related to the impact of HTSs on comfort and performance has been conducted using pilots and military aircrew who are active and required to maintain task vigilance during helicopter flight while wearing the HTS (Ducharme, 2006; Faerevik & Reinertsen, 1998). Faerevik and Reinertsen (2003) found that ambient conditions of 40 °C caused high levels of heat stress (indicated by increases in rectal core temperature, skin temperature, heart rate, and body water loss) and that task vigilance was significantly reduced under these thermal conditions.

With passengers, Taber et al. (2011) observed a mean change in rectal temperature of only 0.08 °C with no decrement in the ability to react to an emergency announcement and complete a helicopter egress skill sequence. This performance was observed after 90 min of exposure to 34 °C while wearing a CAN/CGSB-65.17-99-approved HTS that was unzipped to chest level with the hood removed, as the suit would be worn in flight. It is possible that since participants were expecting an emergency announcement, their level of awareness may have been biased, causing them to mentally rehearse the procedures more than under normal flight conditions.

It is currently unknown how passengers would respond to unexpected emergency announcements or indications of problems with a flight, when their training is not as current and the stresses of a real threat initiate strong sympathetic (fight, flight, or freeze) (Leach, 2004—Chapter 3) nervous system responses (increased heart and breathing rate, increased blood pressure and brain perfusion, for example) (Leach, 1994; Everly & Lating, 2013).

Faerevik and Reinertsen (2003) did report a significant correlation between the increase in core temperature and the number of incorrect reactions that occurred on basic reaction tests. However, the results from Taber et al. (2011) suggested that even though thermal comfort and sensation increased significantly, for flights up to 90 min, physiological heat strain did not occur and there was no reduction in the ability to respond correctly to emergency warnings.

Even if the potentially hot and uncomfortable conditions of helicopter flight do not cause psychological and cognitive impairments, the sweating that occurs due to the heat will reduce the insulation of immersion suits upon immersion. The participants in the Faerevik and Reinertsen (2003) study began to sweat at 40 °C after only 15 min, and the inner and middle clothing layers were saturated after 110 min. Taber et al. (2011) reported the average sweat amount of 0.11 kg after 90 min at 34 °C. These results would suggest that significant wetting of clothing layers, and reduction of thermal insulation, over the duration of a moderate-length flight under warm conditions is likely to occur. Even if physiological heat stress is not induced to the point of cognitive decrement, minimizing the thermal strain on helicopter occupants to improve comfort and reduce sweating is an important aspect of maintaining thermal protection during water entry and should improve the mood and comfort of passengers during transit.

9.10 FLOTATION REQUIREMENTS FOR HELICOPTER PASSENGERS

To minimize buoyancy and meet the EB demands of underwater helicopter egress, buoyancy devices used with HTS systems are always inflatable with a manually activated, compressed carbon dioxide inflation system and an oral inflation backup. Following an underwater egress, the life jacket can be inflated on the surface to increase airway freeboard (i.e., distance between the surface of water and the airways) and improve the flotation position of the wearer.

The life jacket requirements for HTS systems are rooted in regional aviation and airworthiness regulations. These regulations (e.g., FARs, CARs, and JARs), which are referenced by many nations, specify in what helicopter flight situations life jackets must be worn, as well as under which conditions HTS systems are required for thermal protection.

In February 2014, the FAA published guidelines (14 CFR-Part 135, Section 135.168, Subpart 1) for commercial helicopter operations over water, requiring Technical Standard Order (TSO)-C13f-approved life jackets to be worn for flights beyond autorotational distance from shore. There is no mention of HTS use in FARs. This is probably because commercial helicopter flights over cold water are not as common in the United States as they are in colder countries and regions. CARs and JARs both define when helicopter passengers require an HTS (CAR 602.63; JAR-OPS 3.837). When the regulations specify the use of an HTS, the referenced HTS standard will define the flotation performance characteristics of the system.

When the applicable aviation regulations require a life jacket for each passenger, the requirements are commonly met by matching the HTS with an approved aviation inflatable life jacket (see Figure 9.1). In Canada, inflatable buoyancy elements (not a separate, type-approved life jacket—type approval is the term used for the approval of an item to a standard), built to the TSO-C13f life-jacket standard, are often integral to the suit system (see Figures 9.1 and 9.3). The caveat to this approach is that even though the buoyancy element is constructed to the same standard, an integrated buoyancy element that is permanently attached to the suit cannot meet all of the requirements of TSO C13f, which specifies many aspects of how the device must fit and support the body. Thus, when a non-TSO C13f-approved buoyancy element is integrated into an HTS, the aircraft operator will probably need to also carry an approved life jacket for each passenger, even though they are already wearing a superior level of immersion protection.

The TSO- and ETSO-C13f life jacket standards require numerous tests of the material characteristics (e.g., tensile and tear strength) and construction strength (e.g., seam strength), functionality following hot and cold exposures, buoyancy retention, operation forces for the inflator mechanism, donning characteristics, and flotation attitude. After obtaining type approval, a TSO/ETSO-C13f-certified life preserver must then be tested as part of the HTS system type approval to ensure that the required performances defined in the HTS standard are achieved.

9.10.1 Flotation Performance

The flotation characteristics of HTS systems with TSO- or ETSO-C13f-approved life preservers are not always optimal (Armstrong, Bennett-Smith, & Coleshaw, 1994). TSO- or ETSO-C13f-approved life preservers for adults must have at least 155 N (35 lb) of buoyancy, and some life jackets are provided with up to 275 N of buoyancy. The inflated lobes tend to be quite large and can apply pressure to the jaw and neck, obscure vision, and cause difficulty with attaining and maintaining a vertical position and boarding a life raft. One of the most troublesome characteristics is the channeling of waves directly into the airway by the inflated lobes (Armstrong, Bennet-Smith, & Coleshaw, 1994). This channeling of water toward the airway can occur between the lobes of some life jackets that do not have asymmetrically contoured lobes or a system to deflect or block the waves. Without a way to block the waves from channeling toward the airways (potentially even under the spray shield), the risk of drowning is increased.

9.10.2 Spray Shields

Most HTS systems are provided with a spray shield, which is commonly attached to the life jacket, unless an integrated buoyancy element is used. Well-designed spray shields should have the following characteristics: easily deployed, well ventilated, adequate drainage, noncollapsible, and firmly attached for windy conditions. A well-designed spray shield will be effective at protecting the airways (Armstrong et al., 1994; Light & Slater, 1991), increasing survival time (Brooks, 2003; Golden & Tipton, 2002), and reducing water ingress into immersion suits, especially when the suit has a face seal (Sweeney et al., 2008; Sweeney & Taber, 2014).

9.11 SUMMARY

Once a person has survived a helicopter ditching and has reached the surface of the water, he or she is faced with a marine survival situation. An important factor in his or her survival is now related to thermal protection and resources provided by the HTS. The high probability of needing to complete an inverted, underwater egress (Taber, 2014; Taber & McCabe, 2007) to reach the surface introduces a set of performance considerations for HTSs that are not requirements for other types of immersion suits. The purpose of this chapter was to introduce and review how these performance

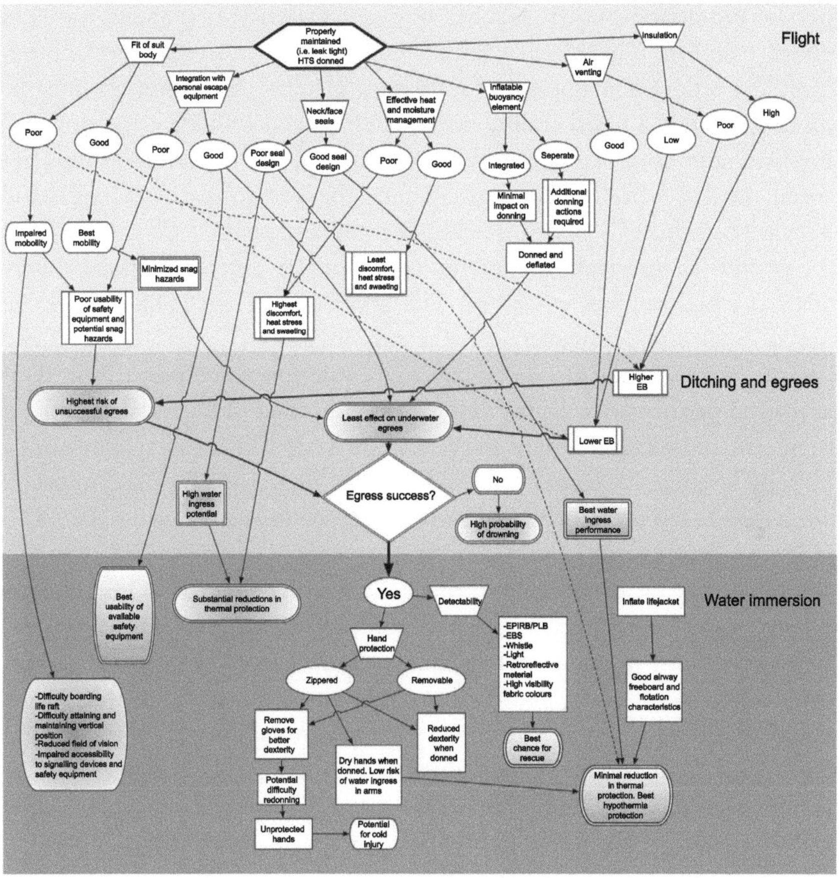

Figure 9.11 Simplified HTS design and performance flowchart showing how the various performance elements must be considered during flight, ditching, and egress and during a prolonged water immersion. Note that some performance variables are considered to affect performance during flight and the water ingress phases, whereas others affect flight and ditching. Key connections are made; however, some are omitted to avoid clutter. Different line styles are used for clarity, not emphasis. Also, some general aspects of immersion suit performance have been omitted from this summary chart and can be found in Figure 2.10 of Sweeney and Taber (2014).

considerations, aviation regulations, and equipment standards constrain and guide the design of HTS systems.

The principal factor that differentiates HTSs from other immersion suits is EB and the constraints it places on other performance elements. EB thresholds are in place to ensure that underwater egress is not unduly impeded by excessive buoyancy. However, HTS systems need to provide thermal protection, which requires buoyant materials or trapped gasses to reduce heat loss. This places EB and thermal protection in direct competition. To reduce the amount of insulation material, water ingress must be minimized with a good sealing system, which can influence the donning simplicity and comfort. After a brief evaluation, the interrelationships between the HTS performance parameters, the performance tradeoffs, and their impact on HTS design quickly become apparent.

As with all survival equipment, there can be distinct regional differences and philosophies of design. The requirements for HTS systems are similar in all cases; however, it is the responsibility of regulators and end users to establish the appropriate balance of performance parameters. This balance of performance priorities is captured in the HTS standards, making it essential to be familiar with the requirements before a suit system can be properly evaluated. Figure 9.11 provides a flowchart of the important aspects of HTS performance that must be considered during the HTS design phase.

Understanding the origins of the standards that lead to HTS products requires an understanding of the philosophies and culture of a region and the stakeholders, in addition to factors such as the environmental conditions, flight transit times and distances, rescue times, and available rescue resources. There will never be one perfect HTS to address all of the various standards, but all of the factors discussed in this chapter frequently get rebalanced with new suit designs that strive to provide the best level of protection and performance for the widely varied environments in which humans travel over cold water in helicopters.

REFERENCES

Allan, J. R. (1984). *Water ingress tests for immersion suits* (No. R504). Royal Air Force-Institute of Aviation Medicine.

Armstrong, I. J., Bennett-Smith, S. C., & Coleshaw, S. R. K. (1994). *Performance of immersion and lifejacket combinations at sea.* Health and Safety Executive. Offshore Technology Report – OTR 94 428.

Barwood, M. J., Dalzell, J., Datta, A. K., Thelwell, R. C., & Tipton, M. J. (2006). Breath-hold performance during cold water immersion: effects of psychological skills training. *Aviation, Space, and Environmental Medicine,* 77, 1136–1142.

Bohemier, A. P., Brooks, C. J., Morton, J. S., & Swain, J. H. (1998). *High fidelity survival training for ditched aircrew and passengers.* Current Aeromedical Issues in Rotary Wing Operations, RTO-MP-19AC/323(HFM)TP/4. Hull, Quebec: Canada Communication Group Inc.

Brooks, C. J. (1988). Maximum acceptable inherent buoyancy limit for aircrew/passenger helicopter immersion suit systems. *Applied Ergonomics, 19*(4), 266–270.

Brooks, C. J. (1989). *The human factors relating to escape and survival from helicopters ditching in water.* RTO AG 305E. Neuilly Sur Seine, AGARD, ISBN 92-835-0522-0.

Brooks, C. J. (2003). *Survival in cold water: Staying alive* (Report TP13822E, 01/2003) Ottawa: Transport Canada.

Brooks, C. J., & Bohemier, A. (1997). Helicopter door and window jettison mechanisms for underwater escape: ergonomic confusion. *Aviation, Space & Environmental Medicine, 68*(9), 844–857.

Brooks, C. J., MacDonald, C. V., Donati, L., & Taber, M. J. (2008). Civilian helicopter accidents into water: analysis of 46 cases, 1979–2006. *Aviation, Space, and Environmental Medicine, 79*, 935–940.

Brooks, C. J., Muir, H. C., & Gibbs, P. N. G. (2001). The basis for the development of a fuselage evacuation time for a ditched helicopter. *Aviation, Space, and Environmental Medicine, 72*, 553–561.

Brooks, C. J., & Provencher, J. D. M. (1984). *Acceptable inherent buoyancy for a ship abandonment/helicopter immersion suit.* DCIEM Report No. 84-C-28.

Canadian Association of Petroleum Producers (CAPP). (2013). *Standard practice for the training and qualifications of personnel.* Report Number 2013–0007.

Canadian Aviation Regulations (CARs). *Section 602.63 life rafts and survival equipment—Flights over water.* Transport Canada. http://laws.justice.gc.ca/eng/regulations/SOR-96-433/page-182.html#h-792.

Canadian General Standards Board. (1988). *Helicopter passenger transportation suit systems CAN/CGSB-65.17-M88.*

Canadian General Standards Board. (1989). *CAN/CGSB-65.16-M89 immersion suit systems standard.* Government of Canada.

Canadian General Standards Board. (1999). *Helicopter passenger transportation suit systems CAN/CGSB-65.17-99.*

Canadian General Standards Board. (2005). *CAN/CGSB-65.16-2005 immersion suit systems standard.* Government of Canada.

Canadian General Standards Board. (2012). *Helicopter passenger transportation suit standard. CAN/CGSB-65.17-2012.*

Cheung, S. S., McLellan, T. M., & Tenaglia, S. (2000). The thermophysiology of uncompensable heat stress. Physiological manipulations and individual characteristics. *Sports Medicine, 29*(5), 329–359.

Civil Aviation Authority. (February 20, 2014). *Safety review of offshore public transport helicopter operations in support of the exploitation of oil and gas.* CAP 1145. http://www.caa.co.uk/application.aspx?catid=33&pagetype=65&appid=11&mode=detail&id=6088.

Coleshaw, S. R. K. (2006). *Investigation of removable exits and windows for helicopter simulators. OPITO Report SC, 153.*

CORD. (1987). *Marine abandonment immersion suit insulation testing parameters: Preliminary report.* Dartmouth, Nova Scotia, Canada: The CORD Group Limited.

CORD. (1995). *Report on the review of the dual role immersion suit systems for Canadian offshore operations.* Report for the National Energy Board of Canada. PERD project number: 6A8001.9.95.

Ducharme, M. B. (2006). Heat stress of helicopter aircrew wearing immersion suit. *Industrial Health, 44*(3), 433–440.

European Aviation Safety Agency (EASA)/ETSO 2C503. (2006). *Helicopter crew and passenger immersion suits.*

European Aviation Safety Agency ETSO-C13f. (2006). *Life preservers.* European technical standard order.

Everly, G. S., & Lating, J. M. (2013). *A clinical guide to the treatment of human stress response.* New York: Springer.

Faerevik, H., & Reinertsen, R. E. (1998). Thermal stress in helicopter pilots, evaluation of two survival suits used during flight. In J. A. Hodgon, J. H. Heaney, & M. J. Buono (Eds.), *Environmental ergonomics VIII* (pp. 173–176). San Diego, CA: Naval Health Research Center, ISBN 0-9666953-1-3.

Faerevik, H., & Reinertsen, R. E. (2003). Effects of wearing aircrew protective clothing on physiological and cognitive responses under various ambient conditions. *Ergonomics, 46*(8), 780–799.

Federal Aviation Administration (FAA). *TSO C13f (1992) Life preservers*. Technical standard order.

Federal Aviation Administration (FAA). (2014). *Helicopter air ambulance, commercial helicopter, and Part 91 helicopter operations-final rule*. Document available https://www.federalregister.gov/articles/2014/02/21/2014-03689/helicopter-air-ambulance-commercial-helicopter-and-part-91-helicopter-operations.

Federal Aviation Regulations (FARs). *14 CFR-Part 135 Section 135.168 emergency equipment: Overwater rotorcraft operations*.

Gagnon, D., McDonald, G. K., Pretorius, T., & Giesbrecht, G. G. (2012). Clothing buoyancy and underwater horizontal swim distance after exiting a submersed vehicle simulator. *Aviation, Space, and Environmental Medicine, 83*, 1077–1083.

Gaul, C. A., & Mekjavic, I. B. (1987). Helicopter pilot suits for offshore application: a survey of thermal comfort and ergonomic design. *Applied Ergonomics, 18*(2), 153–158.

Golden, F. S. C., & Tipton, M. J. (2002). *Essentials of sea survival* (1st ed.). Windsor: Human Kinetics.

Hall, J. J. F., & Polte, J. W. (1956). Effect of water content and compression on clothing insulation. *Journal of Applied Physiology, 8*, 539–545.

Hansen Protection Canada Limited. (2012). *Nautilus HTS-1 suit fitting guide*. Dartmouth, Canada.

Helicopter Safety Advisory Conference (HSAC). (2014). *Helicopter safety advisory conference (HSAC): 2013 gulf of Mexico offshore helicopter operations*. Available from http://www.hsac.org/portals/45/HSAC%202013%20Report.pdf.

Holmes, D. A. (2000). Waterproof breathable fabrics. In A. R. Horrocks, & S. C. Anand (Eds.), *Handbook of technical textiles*. Cambridge, UK: Woodhead Publishing.

International Standards Organization. (2003a). *15027-1 Immersion suits: Constant wear suits, requirements including safety*. Geneva: ISO.

International Standards Organization. (2003b). *15027-3 Test methods: Abandonment suits, requirements including safety*. Geneva: ISO.

Jamieson, D. W., Armstrong, I. J., & Coleshaw, S. R. K. (2001). *Helicopter ditching research: Egress from side-floating helicopters*. CAA Paper 2001/10. London: Civil Aviation Authority.

Jay, O., & White, M. D. (2006). Maximum effort breath-hold times for males and females of similar pulmonary capacities during sudden face-only immersion at water temperatures from 0 to 33 degrees C. *Applied Physiology, Nutrition, and Metabolism, 31*(5), 549–556.

Joint Aviation Authority (JAA). (2007). *Joint aviation regulations (JARs), JAR-OPS 3 commercial air transportation (Helicopters). Section 3.837-Additional requirements for helicopters operating to or from helidecks located in a hostile sea area.*

Kozey, J., Potter, P., & Sweeney, D. (2011). *A comprehensive study of the effects of buoyancy on emergency egress: Phases 1, 2 and 3*. The CORD Group Limited. Unpublished technical report for Petroleum Research Atlantic Canada (PRAC).

Leach, J. (1994). *Survival Psychology*. New York: University Press.

Leach, J. (2004). Why people 'freeze' in an emergency: temporal and cognitive constraints on survival responses. *Aviation, Space, and Environmental Medicine, 75*, 539–542.

Leese, W. L. B., & Norman, J. N. (1979). Helicopter passenger survival suit standards in the UK offshore oil industry. *Aviation, Space, and Environmental Medicine, 50*(2), 110–114.

Light, I. M., Avery, A., & Grieve, A. M. (1987). Immersion suit insulation: the effect of dampening on survival estimates. *Aviation, Space, and Environmental Medicine, 58*(10), 964–969.

Light, I. M., & Slater, P. (1991). *A further in-water performance assessment of lifejacket and immersion suit combinations* (No. OTI 91 550). Aberdeen: RGIT Survival Center Ltd.

MacKinnon, S., & Mallam, S. (2010). *The effect of hand immersion in 4° Celsius water on the performance of helicopter evacuation survival tasks.* Memorial University.

McDonald, G. K., & Giesbrecht, G. G. (2013). Escape from a submersible simulated vehicle simulator wearing different thermoprotective flotation clothing. *Aviation, Space, and Environmental Medicine, 84*(6), 708–715.

Mills, A. M., & Muir, H. (1999). *Development of a standard for underwater survival.* Technical Paper for Shell Group.

Norsk Olje & Gass. (2004). *Norwegian oil and gas recommended guidelines relating to requirement specifications for survival suits for use on the Norwegian continental shelf.* Report-094.

Oil & Gas UK. (2011). *UK offshore commercial air transport helicopter safety record (1981–2010).* The United Kingdom Offshore Oil and Gas Industry Association trading as Oil & Gas UK. Document available at http://www.oilandgasuk.co.uk/cmsfiles/modules/publications/pdfs/HS027.pdf.

OPITO. (2013). *Basic offshore safety induction and emergency training.* Standard Code 5700. Revision 5 (11-Nov 2011) Amendment 4.

Potter, P. L., O'Neill, B., & Brooks, C. J. (2003). *Is it safe to change the CAN/CGSB-65.16-99 marine abandonment immersion suit systems standard to allow a minimum of 70 Newtons of inherent buoyancy?* Dartmouth, Canada: The CORD Group Limited. R03-0004.

Power, J., Simoes-Re, A., Barwood, M., Tikuisis, P., & Tipton, M. (2015). Reduction in predicted survival times in cold water due to wind and waves. *Applied Ergonomics, 49*, 18–24.

Step Change in Safety. (2014a). *Passenger size resources.* https://www.stepchangeinsafety.net/safety-resources/helicopter-safety/helicopter-safety-resources/passenger-size-resources.

Step Change in Safety. (2014b). *Standardized clothing policy.* https://www.stepchangeinsafety.net/news-events/news/standardised-clothing-policy.

Sullivan, P. J., Mekjavik, I. B., & Kakitsuba, N. (1987). Ventilation index of helicopter pilot suits. *Ergonomics, 30*(7), 1053–1061.

Sweeney, D., Ducharme, M. B., Farnworth, B., Prayal-Brown, A., & Potter, P. (2011). The adequacy of 0.75 clo insulation to protect against hypothermia at sea. In *XIV International Conference for Environmental Ergonomics, Nafplio, Greece.*

Sweeney, D., & Potter, P. (2010). *Water ingress in helicopter passenger transportation suits for CAN/CGSB-65.17.* The CORD Group Limited.

Sweeney, D., Potter, P., & Durnford, W. (2008). *Water ingress for immersion suits in Canada: Phase 2.* Dartmouth, Nova Scotia: The CORD Group Limited.

Sweeney, D. H., & Taber, M. J. (2014). Cold water immersion suits. In F. Wang, & C. Gao (Eds.), *Protective clothing: Managing thermal stress.* Cambridge, UK: Woodhead Publishing.

Taber, M. J. (2010). Offshore helicopter safety report. In R. Wells (Ed.), *Offshore helicopter safety inquiry.* St. John's, NL: Canada-Newfoundland and Labrador Offshore Petroleum Board. http://oshsi.nl.ca/?Content=Reports.

Taber, M. J. (2013). Crash attenuating seats: Effects on helicopter underwater escape performance. *Safety Science, 57*, 179–186.

Taber, M. J. (2014). Simulator fidelity and contextual interference in helicopter underwater egress training: an analysis of training and retention of egress skills. *Safety Science, 62*, 271–278.

Taber, M. J., Dies, N. F., & Cheung, S. S. (2011). The effect of transportation suit induced heat stress on helicopter underwater escape preparation and task performance. *Applied Ergonomics, 42*(6), 883–889.

Taber, M. J., MacKinnon, S. N., Power, J., & Walker, R. (2015). Breath-hold times in air compared to breath-hold times during cold water immersions. *Aerospace Medicine and Human Performance, 86*(2), 1–6.

Taber, M. J., & McCabe, J. (2006). Helicopter Ditching: time of crash and survivability. *SAFE Journal, 34*(1), 5–10.

Taber, M. J., & McCabe, J. (2007). An examination of survival rates based on external flotation devices: A helicopter ditching review from 1971 to 2005. *SAFE Journal, 35*(1), 1–6.

Taylor, N. A. S., & Machado-Moreira, C. A. (2013). Regional variations in transepidermal water loss, eccrine sweat gland density, sweat secretion rates and electrolyte composition in resting and exercising humans. *Extreme Physiology and Medicine, 2*, 4.

Tipton, M. J., & Balmi, P. J. (1996). The effect of water leakage on the results obtained from human and thermal manikin tests of immersion protective clothing. *European Journal of Applied Physiology, 72*, 394–400.

Tipton, M. J., & Vincent, M. J. (1989). Protection provided against the initial responses to cold immersion by a partial coverage wet suit. *Aviation, Space, and Environmental Medicine, 60*(8), 769–773.

Transportation Safety Board (TSB) of Canada. (2010). *Aviation investigation report A09A0016. Main gearbox malfunction/Collision with water.* St. John's Newfoundland and Labrador.

CHAPTER 10

Consolidation of Findings and Future Directions for Offshore Helicopter Operations

Michael J. Taber
Falck Safety Services Canada Inc., Dartmouth, NS, Canada

10.1 INTRODUCTION

Since the very first helicopter flights, there have been emergency events that required aircrews to consider what procedures would be needed to evacuate both on land and in the water. Helicopter accidents typically occur in one of three distinct phases of flight: takeoff/landing, hover, and approach (Chapter 2). All three of these phases occur during a routine flight to/from an offshore installation. They also represent the points in the flight that have the highest level of complexity regarding the human/machine interface and operating systems. A malfunction of a system (be it human or machine) may result in an immediate need to land. Research has clearly shown that the chances of survival in a water landing are significantly lower than what would be expected if the event occurred on land (Brooks, MacDonald, Donati, & Taber, 2008; Clifford, 1996; Cunningham, 1978; Taber, 2013, 2014).

This book details the history of helicopter underwater egress training (HUET) from a human systems integration approach in which all influencing factors associated with the environmental conditions, the available technologies, and the individual are considered. The information presented is an attempt to increase our understanding of the factors that influence survival rates in the event of a water impact or ditching. To consider how the information can be consolidated in one collective understanding, the following sections provide a recap/overview of the chapter highlights.

The inclusion of information related to executive functioning should be considered from the perspective of how to better prepare individuals to make decisions in an emergency situation. Leach (Chapter 3) clearly

Handbook of Offshore Helicopter Transport Safety
ISBN 978-1-78242-187-0
http://dx.doi.org/10.1016/B978-1-78242-187-0.00010-9

identifies that there is a limit to what information can be processed during an emergency. Therefore, taking this information into consideration, HUET programs need to be developed with the end user in mind. If the goal is to simply provide a basic introduction to underwater egress, instructional technique and training environment may not need to be at the highest possible level. However, if the goal of a HUET program is to increase the chance of survival following a ditching event, there needs to be a greater emphasis on the critical components that will increase the likelihood that someone can perform the necessary skills during the emergency (Taber, 2014).

10.2 THEORETICAL ESTIMATION OF DITCHING EVENT SURVIVAL

Currently, there is no formalized process for HUET program providers or accident investigation teams to input the data that could be used to inform the development of a theoretical survival curve. In fact, accident investigation teams do not typically record information about previous HUET experience (type of simulator, removal of exit under water, time since last training, seat position) (Brooks et al., 2008). Some aspects such as the influence of equipment, age, sex, anthropometric morphology, hydration level, sleep deprivation, time of day, environmental conditions (e.g., wind, waves, rain, current), and temperature of the water may be included in the investigation report; however, there is no consistency in regard to what is included and what is left out (Brooks et al., 2008).

Understanding the complexities associated with egressing from an inverted and flooded helicopter requires a fundamental examination of factors that may have an impact on survivability. These factors can be separated into two distinct categories (environmental and individual), which, when combined, represent a formidable challenge for those involved in a helicopter ditching. Unfortunately, these challenges are not always overcome and it has been reported that since 2000, a total of 159 offshore personnel have perished in 97 fatal helicopter ditchings (Taber, 2014; see also Baker, Shanahan, Haaland, Brady, & Li, 2011). Similarly, in a review of North Sea ditchings and water impacts, the Civil Aviation Authority (2014) reports that in the 25 reportable events from 1990 to 2012, there were seven fatal accidents with a 29% survival rate, noting that the majority of deaths were the result of drowning. Environmental factors such as inversion of the helicopter (Taber & McCabe, 2007), whether it remains at the surface, water surface conditions, location of equipment (Brooks & Potter, 1998), positioning of seats (Taber, 2013), type of seat restraints, illumination levels, water temperature, presence of petroleum-based products, debris, exit design and placement (Brooks & Bohemier, 1997; Coleshaw, 2006; Taber & Sweeney, 2014), immersion suit design (if applicable—Sweeney & Taber, 2014), and impact forces are only half of the overall survival equation. Individual factors such as age, sex, experience, confidence (Taber & McGarr, 2013), hydration, tolerance to pain, breath-hold capability (Cheung, D'Eon, & Brooks, 2001), fitness level, height, weight, upper body strength, previous injuries, sustained impact injuries, spatial awareness (Cheung, Hofer, Brooks, & Gibbs, 2000), and psychological preparation (Barwood, Dalzell, Datta, Thelwell, & Tipton, 2006) all play a role in survivability. Without exploring each of these factors, in isolation as well as in a combined framework, it is difficult to identify where improvements can be made to overall underwater egress survivability.

10.3 IMPLICATIONS OF HUET TRAINING DIFFERENCES

This book is designed specifically to consider the multitude of factors affecting survival from a high level as well as in detail. Chapter 1 provided a brief overview of why this collection of information is important in the preparation of individuals for a helicopter ditching event. The use of human systems integration to explore how the environment, technology, and the human being influence survival clearly demonstrates that including only one aspect of the event (e.g., training or supplementary survival equipment) will not be sufficient in developing a system that improves survivability. Chapter 1 also began a discussion of legislative requirements for HUET. This discussion is expanded in Chapter 2 to identify the fact that differences in training methodology and the period between refresher courses create tremendous difficulties for recognizing certifications across global offshore jurisdictions.

From a historical perspective, aspects related to HUET for military personnel have been a matter of investigation since World War II. Unfortunately, the typical offshore HUET program has followed a very different path from that of military preparation with regard to training methodology (Chapter 2). This difference is probably based on the belief that military crew members are more often exposed to overwater flight operations and the belief that offshore personnel are not as physically fit as those who serve in the armed forces. Although these differences may be true to a certain extent, the conditions experienced during a ditching are similar regardless of whether the individual is in a military helicopter conducting maritime operations or in a commercial helicopter traveling to an offshore installation. For example, regardless of the occupation (with the exception of pilots), all individuals will experience the same necessary requirements to brace for impact, hold their breath or use an emergency breathing system if the helicopter inverts, egress underwater or abandon to a liferaft, inflate a life vest, activate a personal locator beacon, and wait for rescue. These similarities in task requirements suggest that training should be similar across both groups. Some of the only differences in training should be based on different environmental conditions (e.g., water temperature, time of day, and egress path) that would influence the performance of the tasks. Continued disparity between the training programs will presumably maintain the differences in survival rates for military versus civilian ditching events (Taber, Douglas, & Carroll, 2011).

10.4 COGNITIVE PSYCHOPHYSIOLOGICAL FACTORS INFLUENCING DITCHING SURVIVABILITY

Chapters 3 through 6 of this book focused on the cognitive, psychological, and physiological responses that could influence survival rates following a ditching event. For example, it was noted in Chapter 3 that individuals involved in a time-critical event such as a ditching are bound by a limited number of behaviors that are based on a model of the world around them and not on the actual environment. This difference in the point of reference is extremely important when considering how individuals perform during a ditching event. The potential divide between perception and reality is clearly evident in real-world situations such as that described by one of the survivors of the Shetland Islands ditching, who indicated that several individuals had disconnected their seat belt during or prior to the inversion of the helicopter. Despite the rapid ingress of water as the helicopter was inverting, the fact that individuals prematurely released their harnesses suggests that they initiated a mental plan for a surface abandonment rather than one for an underwater egress.

Chapter 4 provides a foundation for understanding the physiological responses associated with immersion in cold water. By providing an overview of the cold-water shock response, it was possible to make an argument for the inclusion of emergency breathing systems during over-water helicopter operations. Chapter 5 continued and extended the discussion of cold-water shock response to include possible mitigation strategies that significantly influence breath-hold capability. Chapter 6 built on the physiological responses to stressful situations and proposed various psychological interventions that could be employed prior to and during HUET programs.

10.5 FOCUSED DISCUSSION OF FACTORS AFFECTING EGRESS

The information contained in Chapters 7 through 9 focused on specific factors affecting egress. The specificity of the chapters is designed to provide HUET program providers, offshore operators, and workforce personnel with the information necessary to make informed decisions about how to conduct training, what should be considered when recognizing certification from a different jurisdiction, and which program better prepares individuals for an actual ditching event. Furthermore, the specific information highlights the need for individuals to carefully consider the level of detail required to ensure that HUET certification processes and offshore helicopter operations are at an optimal level of safety.

Chapter 7 brought together the current understanding of how fidelity (physical, functional, and cognitive) influences HUET program delivery and possibly the performance of egress skills in a real ditching. The chapter highlighted the specific differences in the types of fidelity to provide a basis for the development of future HUET programs. Given the complexities of performing an underwater egress following a ditching/water impact, it is recommended that any opportunity to safely include a higher level of fidelity in a controlled environment should be considered beneficial. However, an assessment of HUET performance should be carried out to ensure that the increased fidelity improves the probability of survival in a real situation. Understanding the incremental improvements is critical to identifying where, what type, and how much fidelity is desirable.

Chapter 8 explored the inclusion of emergency breathing systems (EBSs) in HUET programs. The information sheds light on the fact that although compressed-air (CA) systems (in various forms) have been used in military HUET programs for over 30 years, the debate about how safe the systems are continued within the offshore community. Despite the concerns of barotrauma during the training of offshore workers, very few events have been documented in the more than three decades of use. Safety in training with a CA EBS is paramount; however, it has been noted that in-air training only presents a risk that individuals will not be adequately prepared to use the system in an actual event (i.e., have not experienced the in-water requirements).

Chapter 9 discussed the requirements for helicopter transportation suits (HTSs). The information regarding suit design and use is intended as a guideline for selecting the most appropriate system for the specific geographical area of use. Maximal egress buoyancy is of particular interest and should be addressed when considering any HTS design. Selection of an appropriate HTS system is a critical component to increasing the probability of survival both inside the helicopter during egress and while waiting for rescue.

10.6 FINAL THOUGHTS AND RECOMMENDATIONS

The purpose of this book is to consolidate the knowledge surrounding our collective understanding of egress and survival following a helicopter ditching/water impact. The ultimate goal of providing this information from some of the world's leading authorities on the subject is to ensure that individuals are better prepared to deal with factors known to be problematic in a ditching, thereby increasing the probability of survival.

From the initial development of a HUET program through to the certification and approval process, there should be a minimum standard that does more than just introduce an overview of what it will take to survive an underwater egress in a real situation. Having a standard that focuses primarily on one aspect (e.g., the reduction of stress) reduces the possibility of fully preparing individuals to perform at an optimal level. For example, the notion that some training is better than no training neglects to take into consideration the fact that if the individual does not have the chance to practice a skill that will probably be required in a real situation, the likelihood of performing the unpracticed skill successfully the first time is limited. Research as well as anecdotal evidence has clearly proven that when individuals are involved in emergency situations with little time to make decisions about processes that they have minimal experience with, the outcome is often not favorable (Brooks et al., 2008; Leach, 2005, 2012; Leach & Ansell, 2008; Leach & Griffith, 2008; Robinson, Sunram-Lea, Leach, & Owen-Lynch, 2008; Taber, 2014). Taken together, the information presented in this book represents an initial resource of information that could be used to develop a HUET program that is better designed to prepare individuals for the tasks that will be necessary to survive a ditching.

Looking forward, it is important for the HUET approval bodies, such as the Canadian Association of Petroleum Producers, OPITO, and Oljearbeidernes Fellessammenslutning (Norwegian Oil Industry Association), to establish standards that focus on how best to train individuals to survive a ditching situation. Reciprocity between international jurisdictions is critical for ensuring that the offshore workforce is trained to a sufficient level. Cross-jurisdictional recognition of certification will allow individuals to travel between work locations without requiring an upgrade. To establish this standard, research will need to be carried out on several aspects of HUET. One of the most important studies will be to better understand the period of HUET skill retention. Currently, there is no process to identify at what point HUET skill performance would fall below an adequate level to aid in survival (Mills & Muir, 1999). To accomplish the goal of a global standard, collaboration between international certification bodies will need to occur. Until the global HUET community has a better understanding of how long HUET skills are retained, it will not be possible to provide any level of prediction of whether a person who was trained six months, one year, two years, three years, or even four years before would be able to complete the tasks that are necessary to egress from a capsized and flooded helicopter.

There have been several calls for HUET to be completed under more realistic conditions (CAA, 2014; Wells, 2010). However, like the standards for HUET program content, there is no regulator process to identify how much fidelity (physical, functional, or cognitive) is appropriate to offer a realistic (but controlled) environment for HUET skill acquisition and retention. For example, there is no standard for how much force should be required to open an underwater egress simulator (UES) exit. This lack of standardization most likely exists because there are no data for the actual situation (Taber & Sweeney, 2014). Although research is currently under way to establish a standard for a Sikorsky S92 in-cabin exit, there are many other helicopters (e.g., AW139, S76, or AS332) being used to transport personnel to offshore installations. At a minimum, all future UESs should be designed to include the actual forces needed to open the exits on the helicopter being used to transport personnel to offshore installations.

These two examples highlight the fact that despite decades of delivering HUET programs to hundreds of thousands of individuals, there are areas of the training and operations that need to be addressed in the coming years. Evidence for this need to consider all aspects of a HUET program is clearly articulated by the lone survivor of the Cougar 491 ditching:

> *A couple days of controlled immersions in a pool every few years is not enough to allow anyone to develop the instinct of reaction that they need to have a chance of escaping a helicopter crash like Cougar 491 (Decker testimony, p. 85—www.oshsi.nl.ca/userfiles/files/HELN5.pdf).*

If operators, training providers, the offshore workforce, and researchers work together, it is reasonable to expect that future HUET programs will better prepare individuals to perform egress skills. Ultimately, this collaboration will increase survival rates in the event of a ditching/water impact.

REFERENCES

Baker, S. P., Shanahan, D. F., Haaland, W., Brady, J. E., & Li, G. (2011). Helicopter crashes related to oil and gas operations in the Gulf of Mexico. *Aviation Space and Environmental Medicine, 82*, 885–889.

Barwood, M. J., Dalzell, J., Datta, A. K., Thelwell, R. C., & Tipton, M. J. (2006). Breath-hold performance during cold water immersion: effects of psychological skills training. *Aviation, Space, and Environmental Medicine, 77*, 1136–1142.

Brooks, C. J., & Bohemier, A. P. (1997). Helicopter door and window jettison mechanisms for underwater escape: ergonomic confusion! *Aviation, Space and Environmental Medicine, 68*, 844–857.

Brooks, C. J., MacDonald, C. V., Donati, L., & Taber, M. J. (2008). Civilian helicopter accidents into water: analysis of 46 cases, 1979–2006. *Aviation, Space and Environmental Medicine, 79*, 35–40.

Brooks, C. J., & Potter, P. L. (1998). The abysmal performance of the inflatable liferaft in helicopter ditchings. In *RTO HFM symposium proceedings*. RTO MP-19.

Cheung, B., Hofer, K., Brooks, C., & Gibbs, P. (2000). Underwater disorientation as induced by two helicopter ditching devices. *Aviation, Space, and Environmental Medicine, 71*(9), 879–888.

Cheung, S. S., D'Eon, N. J., & Brooks, C. J. (2001). Breath-holding ability of offshore workers inadequate to ensure escape from ditched helicopters. *Aviation, Space and Environmental Medicine, 72*, 912–918.

Civil Aviation Authority. (2014). *CAP 1145 – Safety review of offshore public transport helicopter operations in support of the exploitation of oil and gas*. West Sussex, UK: Safety and Airspace Regulation Group.

Clifford, W. S. (1996). *Helicopter crashworthiness. Study I. A review of UK military and world civil helicopter water impacts over the period 1971–1992*. CAA Paper 96005. London: Civil Aviation Authority.

Coleshaw, S. R. K. (2006). Investigation of removable exits and windows for helicopter simulators. In *OPITO report SC 153*.

Cunningham, W. (1978). Helicopter underwater escape trainer (9D5). In *NATO AGARD conference proceedings*, No 255. (Operational Helicopter Aviation Medicine), (pp. 66-1–66-3).

Leach, J. (2005). Cognitive paralysis in an emergency: the role of the supervisory attentional system. *Aviation, Space and Environmental Medicine, 76*, 134–136.

Leach, J. (2012). Maladaptive behaviour in survivors: dysexecutive survivor syndrome. *Aviation, Space and Environmental Medicine, 83*, 1152–1161.

Leach, J., & Ansell, L. (2008). Impairments in attentional processing in a field survival environment. *Applied Cognitive Psychology, 22*, 643–652.

Leach, J., & Griffith, R. (2008). Restrictions in working memory capacity during parachuting: a possible cause of 'no-pull' fatalities. *Applied Cognitive Psychology, 22*, 147–157.

Mills, A. M., & Muir, H. (1999). Development of a standard for underwater survival. In *Technical paper for Shell Group.*

Robinson, S. J., Sunram-Lea, S., Leach, J., & Owen-Lynch, P. J. (2008). The effects of exposure to an acute naturalistic stressor on working memory, state anxiety and salivary cortisol concentration. *International Journal on the Biology of Stress, 11*(2), 115–124.

Sweeney, D. H., & Taber, M. J. (2014). Cold water immersion suits. In F. Wang, & C. Gao (Eds.), *Protective clothing: Managing thermal stress.* Cambridge, UK: Woodhead Publishing.

Taber, M. J. (2013). Crash attenuating seats: effects on helicopter underwater escape performance. *Safety Science, 57*, 179–186.

Taber, M. J. (2014). Simulator fidelity and contextual interference in helicopter underwater egress training—an analysis of training and retention of egress skills. *Safety Science, 62*, 271–278.

Taber, M., Douglas, P., & Carroll, J. (2011). Effect of training on helicopter ditching survival rates. In *European aviation safety agency, helicopter ditching, water impact & survivability conference.*

Taber, M. J., & McCabe, J. (2007). An examination of survival rates based on external flotation devices: a helicopter ditching review from 1971 to 2005. *SAFE Journal, 35*(1), 1–6.

Taber, M. J., & McGarr, G. W. (2013). Confidence in future helicopter underwater egress performance: an examination of training standards. *Safety Science, 60*, 169–175.

Taber, M. J., & Sweeney, D. H. (2014). Forces required to jettison a simulated S92 passenger exit: optimal helicopter underwater egress training techniques. *International Journal of Industrial Ergonomics, 44*, 544–550.

Wells, R. (2010). *Canada-Newfoundland and Labrador offshore helicopter safety inquiry* (Vol. 1). St. John's, NL: Canada-Newfoundland and Labrador Offshore Petroleum Board.

INDEX

Note: 'Page numbers followed by "f" indicate figures and "t" indicate tables.'

U

W

Z

FSC
www.fsc.org
MIX
Papier | Fördert
gute Waldnutzung
FSC® C083411

Kolibri 360 GmbH
Ferdinand-Jühlke-Straße 7
99095 Erfurt, Deutschland
produktsicherheit@kolibri360.de